FIBER OPTIC SENSORS BASED ON PLASMONICS

FIBER OPTIC SENSORS BASED ON PLASMONICS

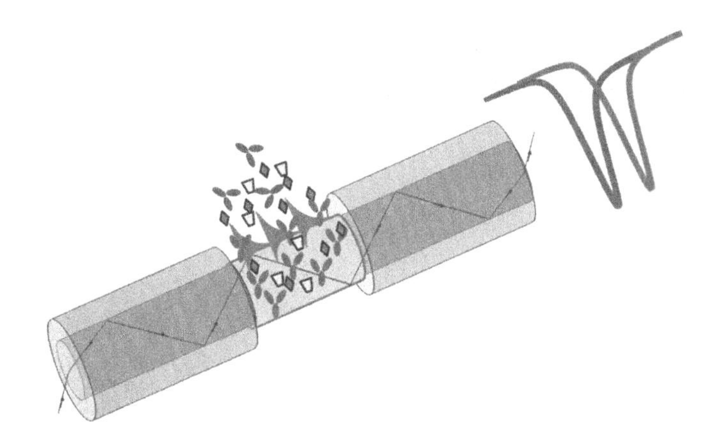

Banshi Dhar Gupta
Indian Institute of Technology Delhi, India

Sachin Kumar Srivastava
Ilse Katz Institute of Nanoscale Science and Technology, Israel

Roli Verma
Indian Institute of Technology Delhi, India

NEW JERSEY • LONDON • SINGAPORE • BEIJING • SHANGHAI • HONG KONG • TAIPEI • CHENNAI

Published by

World Scientific Publishing Co. Pte. Ltd.
5 Toh Tuck Link, Singapore 596224
USA office: 27 Warren Street, Suite 401-402, Hackensack, NJ 07601
UK office: 57 Shelton Street, Covent Garden, London WC2H 9HE

British Library Cataloguing-in-Publication Data
A catalogue record for this book is available from the British Library.

FIBER OPTIC SENSORS BASED ON PLASMONICS

ISBN 978-981-4619-54-7

In-house Editor: Song Yu

Typeset by Stallion Press
Email: enquiries@stallionpress.com

Printed in Singapore

Preface

This book is an effort of bringing a comprehensive book on fiber optic plasmonic sensors for undergraduate and graduate students. The idea resulted due to many enquiries to one of the authors, Banshi D. Gupta, for writing a text book on the subject. Though tremendous amount of literature is available on the research, hardly any text book is available on the subject. Few books have been edited on surface plasmon resonance (SPR)-based sensors, though, to present the advancement of research to the established researchers.

SPR is a phenomenon that can be dealt with a little primary knowledge of electromagnetic theory and hence can be introduced to undergraduate students with its possible applications. A book by Professor Stefan Meir successfully does a part of the job. However, when it comes to the fascinating world of the sensors and their integration with optical fiber, most of the available literature fails to provide a comprehensive collective knowledge of the subject. With this book, we aim to introduce the undergraduate students and early researchers to an overview of the fiber optic sensors based on the fascinating field of plasmonics, their working principles, and applications. The book helps the established researchers with an insight of the current trends in the fiber optic plasmonic sensors.

In the first chapter of the book we have provided a historical overview of the fiber optic sensors based on plasmonics. A chronological development has been presented, along with the remarks at crucial stages of the development. Further, to build the foundation, the Physics of surface plasmons has been discussed in Chapter 2. The chapter comprises of a rigorous discussion and derivation of the wave-vector condition for surface plasmons from Maxwell's equations, issues related to the excitation of surface plasmons, various techniques of excitation, interrogation techniques, and SPR configurations. In the third chapter, we have presented a detailed description of the various components of a sensor, their characteristics, and roles in the sensor. The important characteristics of sensors have also been discussed. In chapter 4, we have presented a rigorous theoretical treatment of fiber optic SPR sensors. This chapter discusses how to incorporate various light launching conditions in optical fiber sensing. Chapter 5 discusses various methods of fabrication of fiber optic SPR sensors. It presents various surface functionalization strategies depending upon the analyte and the medium/ambience of interest. In Chapter 6, we present a number of studies on SPR-based fiber optic sensor using various surface immobilization techniques. Chapter 7 discusses the effect of some of the intrinsic and extrinsic factors, such as temperature, ions, dopants, etc. on the performance of fiber optic SPR sensors. In Chapter 8, we present an overview of the future trend of research and development in the field of fiber optic SPR sensors. We have added two appendices at the end of the book to provide the treatment of the dielectric functions of the metals and tabulated data of various constants and Sellmeier coefficients.

Banshi D. Gupta
Sachin K. Srivastava
Roli Verma

Acknowledgements

I consider myself very fortunate to have worked with Dr. Chandra Deep Singh, Professor Sunil Khijwania and Dr. Navneet K. Sharma on fiber optic evanescent field absorption sensors. My continuous interaction with them has certainly made me wiser in the field of fiber optic sensors. It was this interaction that resulted in the publication of my first book entitled "Fiber optic sensors: Principles and applications (2006). In 2004, I took initiation to work on fiber optic sensors utilizing surface plasmon resonance technique. The research carried out on this topic resulted in 8 Ph.D. theses. I thank Drs. Anuj K. Sharma, Rajan Jha, Rajneesh K. Verma, Sarika Singh, Priya Bhatia, and Satyendra K. Mishra, in addition to Drs. Sachin K. Srivastava and Roli Verma (the other authors of this book), who have worked with me in this exciting field. I learnt a lot from these colleagues through interaction. Indeed, a part of the book has emerged from the work carried out with them. I also thank my present graduate students Rana Tabassum, Anand Mohan Shrivastav and Sruthi Prasood U, for helping me in the preparation of the book. Finally, my special thanks to my wife Uma Gupta for her patience and understanding.

Banshi D. Gupta

Acknowledgements

[illegible]

[illegible]

[illegible]

Contents

Chapter 1

Introduction

1.1 Surface Plasmons: Historical Perspective

The first observation of the phenomenon of surface plasmon resonance (SPR) dates back to 1902, when Wood[1] reported the "*uneven distribution of light in a diffraction grating spectrum*" while he observed patterns of unusual dark and bright bands in the light reflected from a metal backed diffraction grating. Although, he speculated the possible interaction of the grating-metal arrangement with the incident light, no obvious and clear reason for the observed phenomenon was provided.

The first theoretical description of these anomalies was provided by Lord Rayleigh[2] in 1907, when he published the *dynamical theory of gratings*. He assumed that the scattered electromagnetic field can be expanded in terms of outgoing waves only and predicted singularities in the scattered field at several wavelengths. He could predict that these wavelengths (called Rayleigh wavelengths after his name) resemble the Wood's anomalies and occurred only when the electric field was polarized perpendicular to the grating rulings. He called it *s-anomalies*. His theory predicted no singularities for *p-polarization*. However, the existence of *p-anomalies* was reported by Wood in his later observations and further confirmed by Palmer[3,4] when he found the existence of *p-anomalies* in deeply ruled gratings.

Around the same time, in 1907, Zenneck theoretically formulated a special surface wave solution to the Maxwell's equation and demonstrated that the radio frequency surface electromagnetic waves occur at the boundary of two media when one medium is either lossy dielectric or a metal and the other is loss free.[5] He also suggested that the lossy part of the dielectric constant is responsible for the binding of the electromagnetic waves to the interface. Sommerfeld, in 1909, formulated that the field amplitudes of the surface waves postulated by Zenneck, vary inversely as the square root of the distance from the source dipole.[6] In 1941, Fano theoretically concluded that the anomalies reported by Wood[1] in the diffraction grating spectra were observed due to the excitation of surface waves on the interface.[7] Ritchie, in 1957, for the first time coined the word 'surface plasmons' while explaining the characteristic losses of energy experienced by fast electrons when they travel through thin metal films and demonstrated theoretically that the surface plasmons could be excited on the surface of a thin metal film.[8] In 1959, Turbadar reported that illumination of thin metal films on a substrate leads to a large drop in the reflectivity at certain conditions, but this observation was not linked to the excitation of surface plasmons.[9] In 1960, the excitation of surface plasmons at the metal surface was observed by Powell and Swan.[10] They used accelerated electrons for the excitation of surface plasmons. Soon after, in 1960, Stern and Ferrell showed that the electromagnetic waves at the metallic surface involved electromagnetic radiation coupled to the surface plasmons.[11] In 1968, Otto explained the results obtained by Turbadar[9] and demonstrated that in attenuated total internal reflection, the drop in the reflectivity occurred due to the excitation of surface plasmons.[12] Although the first experimental observation of surface plasmons was observed in 1959 by Turbadar, a clear experimental understanding of the phenomenon was presented only in 1968 by Otto.[12] In Otto configuration (discussed briefly in the next section), a small gap of a few nanometers is to be maintained between a prism base and the metal layer, and the dielectric medium is sandwiched between the prism base and the metal layer. The metal layer is coated on a glass substrate. The Otto configuration was not found very suitable from the practical point of view, as it was not easy to maintain an infinitesimal (~100 nm) finite and uniform gap between the metal layer and the prism base. In spite of this, the configuration was very much suitable to the surfaces which would not be damaged or touched by the prism and was important for the study of single crystal surfaces and the roughness of metal films. In 1971, Kretschmann and Raether modified the Otto configuration[13,14] by directly coating the metal film on the base of the prism. In this case, the leaky radiation through the prism–metal interface excites surface plasmons at the metal and outer dielectric medium interface. This search revolutionized the world of surface plasmons. After this, scores of studies were performed to find the effects of prism material, choice of metals and their thickness for efficient coupling of excitation light to surface plasmons. In 1983, Liedberg and co-workers used the phenomenon of surface plasmons first time for sensing applications.[15] They used the basic prism-based Kretschmann configuration and utilized the fact that the resonance wavelength corresponding to the surface plasmons changes with respect to a change in the outer dielectric medium. It was a great step and breakthrough in the direction of use of the surface plasmons for sensing applications. Since then, thousands of studies have been reported on SPR-based sensors. Apart from this, another breakthrough was made in 1993, when Jorgenson and Yee[16] combined the optical fiber technology with surface plasmons by replacing the prism of the Kretschmann configuration by the core of the optical fiber. The collaboration of the optical fiber technology with surface plasmons leads to the miniaturization, cost-effectiveness, and capabilities of online monitoring and remote sensing.

1.2 Kretschmann and Otto Configurations

The Kretschmann configuration based on attenuated total reflection (ATR) is generally used for the excitation of surface plasmons in sensing applications. In this configuration the base of the high refractive index prism is coated with a thin layer of metal and the dielectric medium to be sensed is kept in contact with the other side of the metal layer. To excite surface plasmons the p-polarized light is incident on the prism base through one of its surfaces. The other configuration, Otto configuration, is only used if the roughness of the surface is to be checked because it is difficult to maintain a uniform and nanometer wide gap between the prism base and the metallic surface coated on a glass substrate. The details of these configurations and their use in sensing are discussed in Sec. 2.3.

1.3 Fiber Optic SPR Sensor Developments

The SPR technique has been widely used in sensing of various physical, chemical, and biological parameters/analytes due to simple and reliable procedure, high sensitivity, and fast response. In the case of fiber optic SPR sensor, the prism used in Kretschmann configuration is replaced by the core of an optical fiber. There are many advantages of optical fiber SPR sensor such as miniaturized probe because of a small core diameter allowing a very small quantity of the sensing sample, capability of online monitoring, and remote sensing of the sample. In the fiber optic SPR probe, the metal layer is coated on the unclad core of the fiber and the fluid to be sensed is kept around the metal layer. The concept of combining SPR technique with optical fiber technology has been intensively used for the sensing of various entities such as refractive index of the fluid, film thickness, surface roughness, pH, temperature, urea, glucose, different kinds of pollutants in water and environment, and different kinds of gases. The development of SPR-based fiber optic sensors started in early nineties of last century. The first study on fiber optic chemical sensor based on SPR was reported by Jorgenson and Yee.[16] These authors carried out both the theoretical and experimental studies on the SPR probe fabricated by coating a thin film of metal on an unclad core of a multi-mode optical fiber. A white light source was used to launch the light in the fiber and the spectrum of the transmitted power was recorded for different refractive indices of the fructose solution in water as the sample around the probe. Effect of the sensing length was also studied both theoretically and experimentally and was compared. The refractive index sensitivity and dynamic range are some of the important parameters that define the performance of a sensor. Jorgenson and Yee[17] later investigated these two parameters and found sensitivity to refractive index between 5×10^{-5} and 5×10^{-4} RIU, while the dynamic range was found to be between 1.25 and 1.40 RIU which was tuned from 1.00 to 1.40 RIU by adding a thin high refractive index layer over metal layer. The refractive index range is suitable for both gaseous and aqueous samples. In the same study they reported that the upper limit of the refractive index can be extended up to 1.70 by using sapphire core fiber. Similar kind of fiber optic SPR sensor for refractive index sensing was reported in 1996. The only difference was that instead of a white light source, a monochromatic light source was used for the excitation and the measurements were made at different angles of incidence.[18] The sensor was able to detect refractive index variation as low as 5×10^{-5} in the range 1.35 to 1.40. During the same time, an optical fiber SPR sensor using a side polished single mode fiber with a thin metal over-layer was reported.[19] It was demonstrated that a very small (2×10^{-5}) variation in the refractive index of the sensing sample can be measured by measuring the power transmitted through the fiber. The other advantage of this kind of sensor is that a very small amount of sample is required for the refractive index measurement. In the beginning, silver and gold were used as plasmonic metals for the fabrication of SPR probes. The disadvantage of silver is that it is not very stable and gets oxidized in air. Due to this shortcoming, lots of work on SPR, in the beginning, was carried out using gold coating on prism base/fiber core. To avoid oxidation of silver, self-assembled monolayers (SAMs) of n-alkanethiol was used to cover the silver film coated on the fiber core.[20] The study was extended for the sensing of few percent of chlorofluorocarbon (CFC) vapors by depositing a thin layer of polyfluorosiloxane over SAM.[21]

The side-polished single mode fiber optic SPR sensor reported by Homola and Slavik[19] was modified to achieve high sensitivity.[22] The output end of the fiber was mirrored for back reflecting the light reaching at the output end. Further, the metal layer was coated with pentoxide over-layer. The sensor was able to detect changes in the refractive index lower than 4×10^{-5}. Metal coated tapered fibers with quasi-circular and asymmetric coatings were studied for refractive index sensing.[23] Both types of probes can be used in wavelength and amplitude modes. The quasi-circular probe is polarization insensitive, while the asymmetric probe is polarization sensitive and gives a number of dips corresponding to different modes of surface plasmon. The use of set of dips, improve the measurement accuracy of the refractive index.

In the studies discussed earlier, step index optical fibers were used for SPR probe. A new kind of SPR-based fiber optic sensor was studied both theoretically and experimentally in which an optical fiber with an inverted graded-index profile was used.[24] The sensor exhibited high sensitivity for changes in refractive index of the sensing medium in the range from 1.33 to 1.39. Moreover, palladium was coated on the core of the multi-mode fiber in place of noble metal, in a study, for the sensing of hydrogen gas.[25] The sensor was based on the change in the complex permittivity of palladium in the presence of hydrogen gas. The sensor was able to measure as low as 0.8% concentration of hydrogen in pure nitrogen. The response time varied from 3 s for pure hydrogen to 300 s for its lowest concentrations.

The performance of the SPR-based fiber optic sensor was further improved by using polarization maintaining fiber.[26] The polarization maintaining fibers deliver stable polarization to the SPR probe. The experimentally obtained refractive index resolution of the sensor was better than 4×10^{-6} RIU. In another study carried out to enhance the performance of the sensor, a tapered probe with uniform waist was fabricated on a single mode fiber.[27] The taper was fabricated by stretching a section of fiber during heating with a travelling gas burner. The sensitivity of the sensor was adjustable with the taper waist diameter. Later, in a similar type of study, the uniform waist single mode tapered fiber was coated on one side with a thin metal layer.[28] Due to non-uniform film thickness the SPR spectrum exhibited several dips which depended on the diameter of the waist. The monitoring of multiple dips can improve the accuracy of measurements. During the same time, similar kind of tapered waist was coated with the double layer (metal and dielectric).[29] The coating was asymmetric and was on one side of the cross section of the taper waist. As mentioned earlier, in one of the studies the varying film thickness exhibited multiple resonance dips in SPR curve which increased the dynamic range of the sensor. However, the dielectric layer introduced helped in tuning the response of the sensor to refractive index of the sensing medium. The same sensor based on reflective configuration was later proposed.[30]

In a new approach for a fiber optic SPR sensor two fibers with different core diameters were connected by thermal fusion splicing to leak the transmitted power into the cladding layer of small core diameter fiber so that the leaked light may induce an evanescent wave required for SPR excitation.[31] Additionally, in this structure, there was no need of removing cladding of the fiber. Later, same study was carried out for different thicknesses of the metal layer to obtain the best performance of the sensor and the optimization of the metal film thickness.[32] To operate the hetero-core optical fiber SPR sensor around 1310 nm, a 60 nm thin layer of tantalum pentaoxide (Ta_2O_5) was deposited over metal layer to sense water using a light emitting diode (LED) and intensity interrogation.[33] The advantage of using 1310 nm wavelength is that the scattering loss of silica optical fiber is low. Using this configuration of the probe a multi-point water-detection and water-level gauge devise was designed and tested.

To improve the sensitivity of SPR-based fiber optic sensor, a single mode D-type optical fiber with intensity interrogation method was analyzed.[34] Further, in a different kind of study, a model of absorption-based fiber optic SPR sensor for the measurement of concentration of chemicals was proposed.[35]

In the studies reported earlier, the SPR probe was in the middle of the fiber. After that a fiber optic refractive index microsensor having sensing element at one of the ends of the fiber was developed.[36] The tapered sensing element was coated with gold layer and the white-light source was used for the excitation of surface plasmons. The resonance wavelengths for different refractive indices of the sensing medium were determined from the spectra of the reflected light. A similar type of single mode fiber optic conical microsensor based on SPR was reported for refractive index measurements.[37] The SPR probe at the end of the fiber was also exploited with uniform core but with a layer of silicon oxide (SiO) between fiber core and the metal layer.[38] The addition of SiO layer extended the detection upper limit of refractive index and kept the toughness of the fiber at the same time. A SPR-based fiber optic tip sensor was theoretically analyzed using reflecting spectrum and predicted resonance wavelength and the sensitivity.[39]

Till now in all the studies, only one SPR probe was designed and the refractive index of the liquid or the gas was determined from the location of the dip in the SPR spectrum of the transmitted or reflected light output. In the case of any fluctuations in external environment, the sensitivity and stability of the sensor can change resulting in the wrong interpretation of the measuring parameter. To avoid this, a dual-channel SPR-based fiber optic sensor for biological applications was proposed.[40] The proposed sensor was designed with two SPR sensing elements having different coatings on the same fiber. The SPR spectrum of such a sensor recorded contained two dips corresponding to two sensing elements. The one dip was used as a reference, while the other was used as a signal.

A fiber optic SPR refractive index sensor was analyzed in terms of sensitivity and signal-to-noise ratio for different conditions related to metal layer, optical fiber, and light launching along with the capability of remote sensing.[41] A SPR-based fiber optic sensor for the detection of bittering component (Naringin) was reported.[42] A gel layer with entrapped enzyme was coated over silver coated fiber. The sensor utilized wavelength interrogation and was based on enzymatic reaction. The advantage of enzymatic reaction is that it is analyte specific.

In a new development, a SPR-based fiber optic sensor based on the differential reflectance method was proposed.[43] In this sensor, two light sources of different wavelengths were launched in the fiber probe with reflecting output end face. The reflectance of these two wavelengths, were measured for ethanol solutions of different refractive indices kept around the probe. The difference in reflectance at two wavelengths was found to improve the sensitivity twofold. These measurements are generally carried out at room temperature but it should be remembered that the dielectric constants of fiber core, metal layer, and any other over-layer depend on the temperature of the surrounding medium and hence variation/change in temperature can affect the calibration curve of the SPR-based fiber optic sensor. The influence of temperature on the sensitivity of a SPR-based fiber optic sensor was studied theoretically by incorporating the thermo-optic effects in fiber core, metal, and sensing layers.[44] As mentioned earlier, generally, silver and gold, the noble metals, are used for the fabrication of SPR-based fiber optic sensors. However, other metals have also been tried for SPR sensors. A comparative study of the response curves and properties of the sensors having copper and aluminum layers with gold and silver layers was reported.[45] In this study, the surface characterization revealed the presence of oxide layers on all the films except on gold film. As far as propagation of light inside the fiber is concerned, two kinds of rays, meridional and skew, propagate in the fiber. The fraction of these rays inside the fiber depends on the launching conditions of the light. However, the presence of skew rays reduces the sensitivity of the sensor, which implies that the light launching is very important in a fiber optic SPR sensor.[46] The performance of the fiber optic SPR probe was also improved by adding dopants in the fiber core.[47]

In simulations and experiments on SPR-based fiber optic sensors, the surface of the metallic layer is assumed to be plain which may not be the real situation. In view of this, the effect of surface roughness of the metal layer on the SPR curves, were studied through numerical simulations and an experimental approach.[48] It was reported that the roughness affects the location of SPR dip and hence the resonance wavelength. Another important parameter of the SPR probe is the thickness of the metal film. The effect of gold film thickness on the sensitivity of the multi-mode optical fiber SPR sensor was experimentally investigated using white light source.[49] The gold film thickness of about 65 nm was reported to give precise measurement of resonance wavelength and relatively high sensitivity.

Many attempts have also been made to further improve the sensitivity of the sensor by changing the shape of the fiber optic SPR probe. The shapes considered are tapers of different kinds, uniform core SPR probe sandwiched between two tapers and a

U-shaped fiber optic probe.[50–52] Although, the tapering increases the sensitivity of the sensor but the probe becomes very fragile with the increase in the taper ratio or the decrease in the diameter of the probe. To overcome this problem, a multi-tapered probe of small taper ratio was proposed to enhance the sensitivity of the sensor.[53] In the case of U-shaped probe the sensitivity of the sensor increases with the decrease in the bending radius up to a certain value. A similar kind of SPR sensor utilizing bent region of the single mode optical fiber was reported later.[54] The maximum refractive index resolution of the proposed sensor was estimated to be close to 10^{-8} RIU. Generally, plastic clad silica fibers are used for the fabrication of SPR-based fiber optic sensors. These sensors operate in the visible region and hence communication grade fibers which have communication window around 1.5 μm cannot be used. A study was carried out to show that communication grade fibers can also be used for SPR sensors. In this study, a fiber optic SPR probe was designed using asymmetric doubly deposited tapered fiber to operate in the 1.5 μm region for the range of refractive indices of aqueous fluids.[55] The thicknesses of aluminum and titanium oxide (TiO_2) layers were optimized to achieve this. In the same configuration, a chemical attack on the fiber surface was carried out before the deposition of the double layer on the uniform waist tapered fiber optic SPR probe. The attack increased the roughness of the surface and improved the stability of the deposited layer without compromising the performance of the sensor.[56] The sensitivity of the sensor with the same kind of configuration was further improved by replacing TiO_2 layer with indium nitride in the refractive index range 1.415–1.429.[57] A SPR-based fiber optic sensor coated with Au/Ti film around the taper region was recently reported.[58] The effect of taper angle on the absorption characteristics and SPR sensitivity were investigated. The taper angle was changed by changing the taper length or the taper waist diameter. The sensitivity increased with the increase of the taper angle. The sensitivity of the sensor was also increased by adding a high index layer between analyte and metal layer. The addition of 10 nm thick film of silicon between silver and sensing medium in an optical fiber SPR sensor was shown to increase the sensitivity by two times.[59] In the following study it was observed that doping in silicon is also an important factor. The sensor with n-type silicon layer possesses higher sensitivity than that using p-type silicon layer.[60] Since the refractive index of both types of silicon is same, hence, it appears that the majority charge carriers in silicon play an important role in the sensitivity of the sensor.

Comparison of SPR and localized surface plasmon resonance (LSPR)-based fiber optic sensors was also performed. LSPR refers to metal nanoparticles and the excitation of surface plasmons in nanoparticles occurs when the frequency of incident light is equal to the frequency of collective oscillation of conduction electrons in the nanoparticles. It was found that the SPR sensor has much higher sensitivity than the nanoparticles coated LSPR sensor.[61] A different design of SPR sensor was proposed that uses SPR region at the tip of the single mode long period fiber grating. This sensor possessed high sensitivity and a miniaturized sensing area and was based on intensity interrogation.[62]

Along with the work on the improvement of the sensitivity and the operating range of a fiber optic SPR sensor, the work on the detection of various chemical and biological analytes was in full pace. Several types of fiber optic SPR sensors were reported. A SPR-based fiber optic sensor for the detection of a small percentage of water in ethanol was reported.[63] Generally in the studies carried out on SPR sensors no consideration is given to the presence of ions in the sample. A study carried out on the SPR-based fiber optic sensor with ionic and non-ionic samples revealed that the presence of ions in the sample can influence the resonance wavelength.[64] The shift in resonance wavelength was more for ionic sample than for non-ionic ones of the same refractive index. The shift in the resonance wavelength was found to be proportional to the concentration of ions and was attributed to the interaction of ions with the surface electrons of the metal film. A similar kind of study has been recently reported.[65] A SPR-based plastic optical fiber sensor was developed for the detection of metal ion, Fe (III) using Deferoxamine-Self Assembled Monolyer (DFO-SAM) over gold coated optical fiber.[66] The formation of Fe (III)/DFO complex changes the resonance wavelength. The SPR-based fiber optic sensors were also developed for the detection of low-density lipoprotein,[67] pH,[68,69] low glucose concentration,[70] phenolic compounds,[71] urea,[72] chromium,[73] triacylglycerides,[74] and heavy metal ions.[75] Most of these sensors utilize enzymatic reactions and use gel entrapment technique for the immobilization of enzyme over metal coated fiber probe.

Recently, a fiber optic SPR hydrogen sensor based on wavelength interrogation has been theoretically designed.[76] The sensing structure consists of layers of silver, silica, tungsten oxide, and platinum over unclad core of the fiber. The sensing is based on the chemical reaction of hydrogen gas with tungsten and platinum resulting in the change in their dielectric constants. Further, the use of optimized thicknesses of various layers gave better performance of the sensor. In addition to hydrogen gas sensor, the SPR-based fiber optic sensors for the detection of ammonia[77,78] and hydrogen sulphide[79,80] have also been reported.

The possibility of sensing simultaneously two parameters using a single fiber and detection system was analyzed using different coatings on two unclad regions on a single optical fiber.[81,82] The SPR curves of the sensor obtained show two resonance dips corresponding to two channels and the sensing medium around them. The proposed design can be used for multi-channel and multi-analyte sensing. The study was later extended to three channels of SPR-based fiber optic sensor.[83,84] Recently, simultaneous detection of urea and glucose by SPR-based fiber optic multi-analyte sensor has been reported.[85]

A fiber optic sensor based on SPR induced change in birefringence and intensity was reported using asymmetric bimetal coating on the fiber core.[86] The minimum resolvable refractive index of 5.8×10^{-6} with an operating refractive index range of 0.05 was obtained at 632.8 nm wavelength.

Recently, a number of studies have been carried out on SPR-based optical fiber sensors using molecular imprinting (MIP). MIP is a technique for the creation of binding sites with the memory of size, shape, and functional group of imprinted molecules (target molecules) in a polymer surface. It has been shown to be an efficient way of providing functionalized materials and recognizing the specific molecules in a mixture of related compounds. Some of the studies carried out using MIP technique are on the detection of trinitrotoluene,[87] vitamin B_3,[88] tetracycline,[89] and L-nicotine.[90] In addition to MIP-based fiber optic SPR sensors, considerable interest has also been shown on the use of graphene layer in SPR probe configuration. It has been reported that the addition of a graphene layer over metal layer enhances the sensitivity of the SPR sensor. The studies carried out recently include the detection of refractive index[91] and ammonia gas.[92]

1.4 Overview of the Book

From the last section it is clear that a tremendous amount of work on plasmonics-based fiber optic sensors whether it is on the enhancement of sensitivity or operating range of the sensor or on the development of sensors for different analytes is going on in different research groups all over the world. This is because of the number of advantages of these kinds of sensors. The principle of the SPR-based sensor relies on the change in the refractive index of the sensing medium/layer around the metal layer. The simplest SPR sensor is the one in which the sensing medium (a fluid) is kept around the metal layer and the change in its refractive index changes the resonance wavelength in the SPR spectrum. This can be given the name as direct SPR sensor. In another class of SPR sensor, a layer of dielectric medium with some modification is coated over the metal layer and its refractive index changes when the sensing fluid comes in its contact. The change in its refractive index can be due to (i) enzymatic reaction between enzyme embedded layer and the corresponding analyte present in the fluid, or (ii) swelling/shrinkage of gel layer due to the analyte, or (iii) attachment of analyte in the pores created for it using MIP technique. This kind of sensor can be named as indirect SPR sensor. The present book consists of eight chapters including the present one in which we have given the historical perspective of surface plasmons and have presented the development of the field since 1993. In

Chapter 2, we shall develop the field by introducing the surface plasmons for semi-infinite metal–dielectric interface and shall define their propagation length and penetration depth. Practical issues with the excitation of surface plasmons in different configurations and in various geometries will be discussed. We shall then introduce some specially named SPR geometries of recent interests. Finally, we shall present various means of interrogation of SPR and shall end up with a note on SPR imaging. In Chapter 3, we shall introduce optical fiber and shall discuss the principle and designs of common fiber optic sensors, their components, functions, and the performance parameters. Chapter 4 will present the theoretical description of fiber optic SPR sensors with respect to various light launching conditions and the performance parameters. This will be followed by the fabrication and functionalization methods and protocols used for the fabrication of fiber optic SPR biosensors in Chapter 5. In Chapter 6, we shall discuss, in detail, some of the fiber optic sensing applications based on SPR phenomena and mentioned briefly in the previous section. Various issues, such as sensitivity enhancement, influence of external stimuli, etc. will follow in Chapter 7. The last chapter of the book will summarize all the contents with a note on future trends of research in the field of fiber optic sensors based plasmonics.

References

1. R.M. Wood, On a remarkable case of uneven distribution of light in a diffraction grating spectrum, *Philosophical Magazine* **4** (1902) 396–402.
2. L. Rayleigh, On the dynamical theory of gratings. *Proceedings of Royal Soc. London* **79** (1907) 399–416.
3. C.H. Palmer, Parallel diffraction grating anomalies, *Journal of Optical Society of America* **42** (1952) 269–273.
4. C.H. Palmer, Diffraction grating anomalies. II. Coarse gratings, *Journal of Optical Society of America* **46** (1956) 50–53.
5. J. Zenneck, Uber die Fortpflanztmg ebener elektro-magnetischer Wellen langs einer ebenen Leiterflache und ihre Beziehung zur drahtlosen Telegraphie, *Annals der Physik* **23** (1907) 846–866.
6. A. Sommerfeld, Propagation of waves in wireless telegraphy, *Annals der Physik* **28** (1909) 665–736.

7. U. Fano, The theory of anomalous diffraction gratings and of quasi-stationary waves on metallic surfaces (Sommerfeld's waves), *Journal of Optical Society of America* **31** (1941) 213–222.
8. R.H. Ritchie, Plasma losses by fast electrons in thin films, *Physical Review* **106** (1957) 874–881.
9. T. Turbadar, Complete absorption of light by thin metal films, *Proc. Phys. Soc.* **73** (1959) 40–44.
10. C.J. Powell and J.B. Swan, Effect of oxidation on the characteristic loss spectra of aluminum and magnesium. *Physical Review* **118** (1960) 640–643.
11. E.A. Stern and R.A. Ferrell, Surface plasma oscillations of a degenerate electron gas. Physical Review **120** 130–136 (1960).
12. A. Otto, Excitation of nonradiative surface plasma waves in silver by the method of frustrated total reflection, *Zeitschrift für Physik* **216** (1968) 398–410.
13. E. Kretschmann and H. Raether, Radiative decay of non-radiative plasmons excited by light, *Z. Naturforsch*, **A23** (1968) 2135–2136.
14. E. Kretschmann, The determination of the optical constants of metals by excitation of surface plasmons. *Z. Physik* **241** (1971) 313–324.
15. B. Liedberg, C. Nylander and I. Lundstrom, Surface plasmon resonance for gas detection and biosensing, *Sensors and Actuators* **4** (1983) 299–304.
16. R.C. Jorgenson and S.S. Yee, A fiber-optic chemical sensor based on surface plasmon resonance, *Sensors and Actuators B* **12** (1993) 213–220.
17. R.C. Jorgenson and S.S. Yee, Control of the dynamic range and sensitivity of a surface plasmon resonance based fiber optic sensor, *Sensors and Actuators A* **43** (1994) 44–48.
18. A. Trouillet, C. Ronot-Trioli, C. Veillas and H. Gagnaire, Chemical sensing by surface plasmon resonance in a multimode optical fiber. *Pure and Applied Optics* **5** (1996) 227–237.
19. J. Homola and R. Slavik, Fibre-optic sensor based on surface plasmon resonance, *Electronics Letters* **32** (1996) 480–482.
20. A. Abdelghani, J. M. Chovelon, J. M. Krafft, N. Jaffrezic-Renault, A. Trouillet, C. Veillas, C. Ronot-Trioli and H. Gagnaire, Study of self-assembled monolayers of n-alkanethiol on a surface plasmon resonance fiber optic sensor, *Thin Solid Films* (1996) 284–285, 157–161.
21. A. Abdelghani, J.M. Chovelon, N. Jaffrezic-Renault, C. Ronot-Trioli, C. Veillas and H. Gagnaire, Surface plasmon resonance fiber-optic sensor for gas detection, *Sensors and Actuators B* **39** (1997) 407–410.
22. R. Slavík, J. Homola, and J. Ctyroký, Miniaturization of fiber optic surface plasmon resonance sensor, *Sensors and Actuators B* **51** (1998) 311–315.
23. A. Díez, M.V. Andrés and J.L. Cruz, In-line fiber-optic sensors based on the excitation of surface plasma modes in metal-coated tapered fibers, *Sensors and Actuators B* **73** (2001) 95–99.
24. F. Bardin, I. Kasik, A. Trouillet, V. Matejec, H. Gagnaire and M. Chomat, Surface plasmon resonance sensor using an optical fiber with an inverted graded-index profile, *Applied Optics* **41** (2002) 2514–2520.

25. X Bévenot, A Trouillet, C Veillas, H Gagnaire and M Clément, Surface plasmon resonance hydrogen sensor using an optical fiber, *Measurement Science and Technology* **13** (2002) 118–124.
26. M. Piliarik, J. Homola, Z. Maníková and J. Ctyroký, Surface plasmon resonance sensor based on a single-mode polarization-maintaining optical fiber, *Sensors and Actuators B* **90** (2003) 236–242.
27. J. Villatoro, D. Monzon-Hernandez and E. Mejia, Fabrication and modeling of uniform-waist single-mode tapered optical fiber sensors, *Applied Optics* **42** (2003) 2278–2283.
28. D. Monzón-Hernández, J. Villatoro, D. Talavera and D. Luna-Moreno, Optical-fiber surface-plasmon resonance sensor with multiple resonance peaks, *Applied Optics* **43** (2004) 1216–1220.
29. F.J. Bueno, O. Esteban, N. Diaz-Herrera, M.C. Navarrete and A. Gonzalez-Cano, Sensing properties of asymmetric double-layer-covered tapered fibers, *Applied Optics* **43** (2004) 1615–1620.
30. O. Esteban, N. Diaz-Herrera, M.C. Navarrete and A. Gonzalez-Cano, Surface plasmon resonance sensors based on uniform-waist tapered fibers in a reflective configuration, *Applied Optics* **45** (2006) 7294–7298.
31. M. Iga, A. Seki and K. Watanabe, Hetero-core structured fiber optic surface plasmon resonance sensor with silver film, *Sensors and Actuators B* **101** (2004) 368–372.
32. M. Iga, A. Seki and K. Watanabe, Gold thickness dependence of SPR-based hetero-core structured optical fiber sensor, *Sensors and Actuators B* **106** (2005) 363–368.
33. K. Takagi and K. Watanabe, Near infrared characterization of hetero-core optical fiber SPR sensors coated with Ta_2O_5 film and their applications, *Sensors* **12** (2012) 2208–2218.
34. M.H. Chiu, C.H. Shih and M.H. Chi, Optimum sensitivity of single-mode D-type optical fiber sensor in the intensity measurement, *Sensors and Actuators B* **123** (2007) 1120–1124.
35. A.K. Sharma and B.D. Gupta, Absorption-based fiber optic surface plasmon resonance sensor: A theoretical evaluation, *Sensors and Actuators B* **100** (2004) 423–431.
36. B. Grunwald and G. Holst, Fiber optic refractive index microsensor based on white-light SPR excitation, *Sensors and Actuators A* **113** (2004) 174–180.
37. K. Kurihara, H. Ohkawa, Y. Iwasaki, O. Niwa, T. Tobita and K. Suzuki, Fiber-optic conical microsensors for surface plasmon resonance using chemically etched single-mode fiber, *Analytica Chimica Acta* **523** (2004) 165–170.
38. J. Zeng and D. Liang, Application of fiber optic surface plasmon resonance sensor for measuring liquid refractive index, *Journal of Intelligent Material Systems and Structures* **17** (2006) 787–791.
39. Y. Yuan, D. Hu, L. Hua and M. Li, Theoretical investigations for surface plasmon resonance based optical fiber tip sensor, *Sensors and Actuators B* **188** (2013) 757–760.

40. W. Peng, S. Banerji, Y.C. Kim and K.S. Booksh, Investigation of dual-channel fiber-optic surface plasmon resonance sensing for biological applications, *Optics Letters* **30** (2005) 2988–2990.
41. A.K. Sharma and B.D. Gupta, On the sensitivity and signal-to-noise ratio of a step-index fiber optic surface plasmon resonance sensor with bimetallic layers, *Optics Communications* **45** (2005) 159–169.
42. Rajan, S. Chand and B.D. Gupta, Fabrication and characterization of a surface plasmon resonance based fiber-optic sensor for bittering component-naringin, *Sensors and Actuators B* **115** (2006) 344–348.
43. H. Suzuki, M. Sugimoto, Y. Matsui and J. Kondoh, Fundamental characteristics of a dual-colour fiber optic SPR sensor, *Measurement Science and Technology* **17** (2006) 1547–1552.
44. A.K. Sharma and B.D. Gupta, Influence of temperature on the sensitivity and signal-to-noise ratio of a fiber optic surface plasmon resonance sensor, *Applied Optics* **45** (2006) 151–161.
45. M. Mitsushio, K. Miyashita and M. Higo, Sensor properties and surface characterization of the metal-deposited SPR optical fiber sensors with Au, Ag, Cu, and Al, *Sensors and Actuators A* **125** (2006) 296–303.
46. Y.S. Dwivedi, A.K. Sharma and B.D. Gupta, Influence of skew rays on the sensitivity and signal to noise ratio of a fiber-optic surface-plasmon-resonance sensor: A theoretical study, *Applied Optics* **46** (2007) 4563–4569.
47. A.K. Sharma, Rajan and B.D. Gupta, Influence of different dopants on the performance of a fiber optic SPR sensor, *Optics Communications* **274** (2007) 320–326.
48. M. Kanso, S. Cuenot and G. Louarn, Roughness effect on the SPR measurements for an optical fiber configuration: experimental and numerical approaches, *Journal of Optics A: Pure and Applied Optics* **9** (2007) 586–592.
49. H. Suzuki, M. Sugimoto, Y. Matsui and J. Kondoh, Effects of gold film thickness on spectrum profile and sensitivity of multimode-optical-fiber SPR sensor, *Sensors and Actuators B* **132** (2008) 26–33.
50. R.K. Verma, A.K. Sharma and B.D. Gupta, Modeling of tapered fiber-optic surface plasmon resonance sensor with enhanced sensitivity, *IEEE Photonics Technology Letters* **19** (2007) 1786–1788.
51. R.K. Verma, A.K. Sharma and B.D. Gupta, Surface plasmon resonance based tapered fiber optic sensor with different taper profiles, *Optics Communications* **281** (2008) 1486–1491.
52. R.K. Verma and B.D. Gupta, Theoretical modelling of a bi-dimensional U-shaped surface plasmon resonance based fiber optic sensor for sensitivity enhancement, *Journal of Physics D: Applied Physics* **41** (2008) 095106.
53. S.K. Srivastava and B.D. Gupta, A multi-tapered fiber-optic SPR sensor with enhanced sensitivity, *IEEE Photonics Technology Letters* **23** (2011) 923–925.
54. Y.N. Kulchin, O.B. Vitrik, A.V. Dyshlyuk and Z. Zhou, Conditions for surface plasmon resonance excitation by whispering gallery modes in a bent single mode optical fiber for the development of novel refractometric sensors, *Laser Physics* **23** (2013) 085105.

55. N. Díaz-Herrera, A. González-Cano, D. Viegas, J.L. Santos and M.C. Navarrete, Refractive index sensing of aqueous media based on plasmonic resonance in tapered optical fibres operating in the 1.5 μm region, *Sensors and Actuators B* **146** (2010) 195–198.
56. N. Díaz-Herrera, O. Esteban, M.C. Navarrete, A. González-Cano, E. Benito-Pena and G. Orellana, Improved performance of SPR sensor by a chemical etching of tapered optical fibers, *Optics and Lasers in Engineering* **49** (2011) 1065–1068.
57. O. Esteban, F.B. Naranjo, N. Díaz-Herrera, S. Valdueza-Felip, M.C. Navarrete and A. González-Cano, High-sensitive SPR sensing with indium nitride as a dielectric overlay of optical fibers, *Sensors and Actuators B* **158** (2011) 372–376.
58. S. Ju, S. Jeong, Y. Kim, P. Jeon, M.S. Park, H. Jeong, S. Boo, J.H. Jang and W.T. Han, Experimental demonstration of surface plasmon resonance enhancement of the tapered optical fiber coated with Au/Ti thin film, *Journal of Non-Crystalline Solids* **383** (2014) 146–152.
59. P. Bhatia and B.D. Gupta, Surface-plasmon-resonance-based fiber-optic refractive index sensor; sensitivity enhancement, *Applied Optics* **50** (2011) 2032–2036.
60. P. Bhatia and B.D. Gupta, Surface plasmon resonance based fiber optic refractive index sensor utilizing silicon layer: effect of doping, *Optics Communications* **286** (2013) 171–175.
61. J. Cao, E.K. Galbraith, T. Sun and K.T.V. Grattan, Comparison of surface plasmon resonance and localized surface plasmon resonance-based optical fiber sensors, *Journal of Physics: Conference Series* **307** (2011) 012050.
62. T. Schuster, R. Herschel, N. Neumann and C.G. Schaffer, Miniaturized long-period fiber grating assisted surface plasmon resonance sensor, *IEEE Journal of Lightwave Technology* **30** (2012) 1003–1008.
63. S.K. Srivastava, R. Verma and B.D. Gupta, Surface plasmon resonance based fiber optic sensor for the detection of low water content in ethanol, *Sensors and Actuators B* **153** (2011) 194–198.
64. S.K. Srivastava and B.D. Gupta, Effect of ions on surface plasmon resonance based fiber optic sensor, *Sensors and Actuators B* **156** (2011) 559–562.
65. L. Feng, W. Zhang, D. Liang and J. Lee, Total dissolved solids estimation with a fiber optic sensor of surface plasmon resonance, *Optik* **125** (2014) 3337–3343.
66. N. Cennamo, G. Alberti, M. Pesavento, G. D'Agostino, F. Quattrini, R. Biesuz and L. Zeni, A simple small size and low cost sensor based on surface plasmon resonance for selective detection of Fe(III), *Sensors* **14** (2014) 4657–4671.
67. R. Verma, S.K. Srivastava and B.D. Gupta, Surface-plasmon-resonance-based fiber-optic sensor for the detection of low-density lipoprotein, *IEEE Sensors Journal* **12** (2012) 3460–3466.
68. S. Singh and B.D. Gupta, Fabrication and characterization of a highly sensitive surface plasmon resonance based fiber optic pH sensor utilizing

high index layer and smart hydrogel, *Sensors and Actuators B* **173** (2012) 268–273.

69. S.K. Mishra and B.D. Gupta, Surface plasmon resonance based fiber optic pH sensor utilizing Ag/ITO/Al/hydrogel layers, *Analyst* **138** (2013) 2640–2646.
70. S. Singh and B.D. Gupta, Fabrication and characterization of a surface plasmon resonance based fiber optic sensor using gel entrapment technique for detection of low glucose concentration, *Sensors and Actuators B* **177** (2013) 589–595.
71. S. Singh, S.K. Mishra and B.D. Gupta, SPR based fiber optic biosensor for phenolic compounds using immobilization of tyrosinase in polyacrylamide gel, *Sensors and Actuators B* **186** (2013) 388–395.
72. P. Bhatia and B.D. Gupta, Fabrication and characterization of a surface plasmon resonance based fiber optic urea sensor for biomedical applications, *Sensors and Actuators B* **161** (2011) 434–438.
73. S.K. Mishra and B.D. Gupta, Surface plasmon resonance based fiber optic sensor for the detection of CrO_4^{2-} using Ag/ITO/hydrogel layers, *Analytical Methods* **6** (2014) 5191–5197.
74. A. Baliyan, P. Bhatia, B.D. Gupta, E.K. Sharma, A. Kumari and R. Gupta, Surface plasmon resonance based fiber optic sensor for the detection of triacylglycerides using gel entrapment technique, *Sensors and Actuators B* **188** (2013) 917–922.
75. R. Verma and B.D. Gupta, Detection of heavy metal ions in contaminated water by surface plasmon resonance based optical fiber sensor using conducting polymer and chitosan, Food Chemistry **166** (2015) 568–575.
76. X. Wang, Y. Tang, C. Zhou and B. Liao, Design and optimization of the optical fiber surface plasmon resonance hydrogen sensor based on wavelength modulation, *Optics Communications* (2013) 298–299, 88–94.
77. S.K. Mishra, D. Kumari and B.D. Gupta, Surface plasmon resonance based fiber optic ammonia gas sensor using ITO and polyaniline, *Sensors and Actuators B* (2012) 171–172, 976–983.
78. P. Bhatia and B.D. Gupta, Surface plasmon resonance based fiber optic ammonia sensor utilizing bromocresol purple, *Plasmonics* **8** (2013) 779–784.
79. R. Tabassum, S.K. Mishra and B.D. Gupta, Surface plasmon resonance-based fiber optic hydrogen sulphide gas sensor utilizing Cu-ZnO thin films, *Physical Chemistry Chemical Physics* **15** (2013) 11868–11874.
80. S.K. Mishra, S. Rani and B.D. Gupta, Surface plasmon resonance based fiber optic hydrogen sulphide gas sensor utilizing nickel oxide doped ITO thin film, *Sensors and Actuators B* **195** (2014) 215–222.
81. Y. Yuan, L. Wang and J. Huang, Theoretical investigation for two cascaded SPR fiber optic sensors, *Sensors and Actuators B* **161** (2012) 269–273.
82. R. Verma, S.K. Srivastava and B.D. Gupta, Surface plasmon resonance based multi-channel and multi-analyte fiber optic sensor, *Proceedings of SPIE* 8351, 83512D (2012).
83. R. Verma and B.D. Gupta, Fiber optic surface plasmon resonance-based three channels multi-analyte sensor, *Chemical Sensors* **3** (2013) 13.

84. R. Verma and B.D. Gupta, SPR based three channels fiber optic sensor for aqueous environment, *Proceedings of SPIE* 8992 (2014) 899209.
85. R. Verma and B.D. Gupta, A novel approach for simultaneous sensing of urea and glucose by SPR based optical fiber multianalyte sensor, *Analyst* **139** (2014) 1449–1455.
86. T.T. Nguyen, E.C. Lee and H. Ju, Bimetal coated optical fiber sensors based on surface plasmon resonance induced change in birefringence and intensity, *Optics Express* **22** (2014) 5590–5598.
87. N. Cennamo, G. D'Agostino, R. Galatus, L. Bibbo, M. Pesavento and L. Zeni, Sensors based on surface plasmon resonance in a plastic optical fiber for the detection of trinitrotoluene, *Sensors and Actuators B* **188** (2013) 221–226.
88. R. Verma and B.D. Gupta, Fiber optic SPR sensor for the detection of 3-pyridinecarboxamide (vitamin B_3) using molecularly imprinted hydrogel, *Sensors and Actuators B* **177** (2013) 279–285.
89. R. Verma and B.D. Gupta, Optical fiber sensor for the detection of tetracycline using surface plasmon resonance and molecular imprinting, *Analyst* **138** (2013) 7254–7263.
90. N. Cennamo, G. D'Agostino, M. Pesavento and L. Zeni, High selectivity and sensitivity sensor based on MIP and SPR in tapered plastic optical fibers for the detection of L-nicotine, *Sensors and Actuators B* **191** (2014) 529–536.
91. J.A. Kim, T. Hwang, S.R. Dugasani, R. Amin, A. Kulkarni, S.H. Park and T. Kim, Graphene based fiber optic surface plasmon resonance for bio-chemical sensor applications, *Sensors and Actuators B* **187** (2013) 426–433.
92. S.K. Mishra, S.N. Tripathi, V. Choudhary and B.D. Gupta, SPR based fiber optic ammonia gas sensor utilizing nanocomposite film of PMMA/reduced graphene oxide prepared by *in situ* polymerization, *Sensors and Actuators B* **199** (2014) 190–200.

Chapter 2

Physics of Plasmons

2.1 Introduction

Nobel metals possess a dense assembly of negatively charged free electrons (free electron charge density $\sim 10^{23}\,\mathrm{cm}^{-3}$) in an equally charged positive ion background. The ensemble of the free electrons and the positive ions can be compared with plasma of particles, as it possesses a highly dense assembly of charged particles with the condition of quasi-neutrality.[1] Since, positive ions possess an infinitely large mass as compared to that of free electrons, therefore, according to the Jellium model, these ions can be termed as a perpetually constant positive background in which free electrons can easily move from one point to the other in such a fashion that the condition for quasi-neutrality is always fulfilled. To say, the total charge density inside the metal always remains zero. If we forget for a while about the constant positive background, a metal can be viewed as a sea of free electrons. Now, we shall analogically understand the phenomenon of plasmons with the waves in a water body. What happens when we drop a pebble in calm water? If we focus on the bulk medium and not the surface, it is observed that a longitudinal wave gets generated which travels away from the position of the disturbance. In a similar fashion, if an external field is applied on a point in the metal, the local density of free electrons at that place in the metal gets changed due to the force of the applied field. Say, for example, the electron density at that point gets locally increased due to the movement of free electrons under the applied field. The movement of the free electrons creates behind a positive center of the background which is no longer screened by the free electrons. The locally dense free electrons begin to get attracted by this unscreened positive ion background. At the same time, the Coulomb repulsion among the moving free electrons also comes into the picture. This attraction acts as a driving force for free electrons and they move to the positive region and accumulate with a density greater than necessary to obtain charge neutrality. Now, at this point, a restoring force produces motion in an opposite direction. The resultant of the two forces, (i.e., attractive driving force and repulsive restoring force) sets up the longitudinal oscillations among the free electrons. These oscillations are known as plasma oscillations. A plasmon is a quantum of the plasma oscillation. The existence of plasma oscillations has been demonstrated in electron energy-loss experiments.[2,3]

A metal–dielectric interface supports plasma oscillations. These charge density oscillations along the metal–dielectric interface are known as surface plasma oscillations. The quantum of these oscillations is referred to as surface plasmon (SP) (also a SP wave or a SP mode). The SPs are accompanied by a longitudinal (TM-polarized) electric field, which decays exponentially in metal as well as in dielectric medium. Due to this exponential decay of field intensity, the field has its maximum at the metal–dielectric interface itself. Both of these crucial properties of SPs being TM-polarized and exponential decay of electric field are found by solving the Maxwell's equations.

In this chapter we shall learn the physics of SPs using Maxwell's equations and boundary conditions in addition to the methods of their excitations. Different kinds of interrogation techniques have been used to apply SP technique for sensing. We shall elaborate these in the later part of this chapter in addition to surface plasmon resonance imaging (SPRI).

2.2 SPs at Semi-Infinite Metal–Dielectric Interface

Consider a metal–dielectric interface as shown in Fig. 2.1. Let the dielectric constants [=(refractive index)^2] of the dielectric medium be ε_s and that of the metal be $\varepsilon_m = \varepsilon_{m1} + i\varepsilon_{m2}$, where ε_{m1} and ε_{m2} are respectively the real and imaginary parts of the metal dielectric function. The associated coordinate system has X-axis along the interface, while the Z-axis is normal to the interface. The Y-axis is normal to the plane of the paper and is out of the plane towards the reader. We first apply the boundary conditions on the solutions of the four Maxwell's equations for both the electric and magnetic field components. Then from the modal fields supported by the interface, we deduce the expressions for propagation constant, propagation length, and penetration depth. The Maxwell's equations are given as:

$$\nabla \cdot \mathbf{D} = \rho_{\text{ext}}, \tag{2.1}$$

$$\nabla \cdot \mathbf{B} = 0, \tag{2.2}$$

$$\nabla \times \mathbf{E} = -\frac{\partial \mathbf{B}}{\partial t}, \tag{2.3}$$

$$\nabla \times \mathbf{H} = \mathbf{J}_{\text{ext}} + \frac{\partial \mathbf{D}}{\partial t}, \tag{2.4}$$

where **D**, **B**, **E** and **H** represent the four macroscopic fields; namely dielectric displacement, magnetic induction, electric field, and magnetic field respectively; while ρ_{ext} and $\mathbf{J}_{\text{ext}}$ represent the external charge and current densities, respectively.

For time varying harmonic fields ($\partial/\partial t = -i\omega$), the four Maxwell's equations can be rearranged to give two sets of self-consistent solutions with different polarization properties. One set of solutions is called transverse magnetic (TM or p) mode and the other set is called transverse electric (TE or s) mode. For the kind of geometry shown in Fig. 2.1, the set of solutions E_x, H_y, and E_z correspond to TM modes and H_x, E_y, and H_z correspond to TE modes. The corresponding TM and TE wave equations are given by

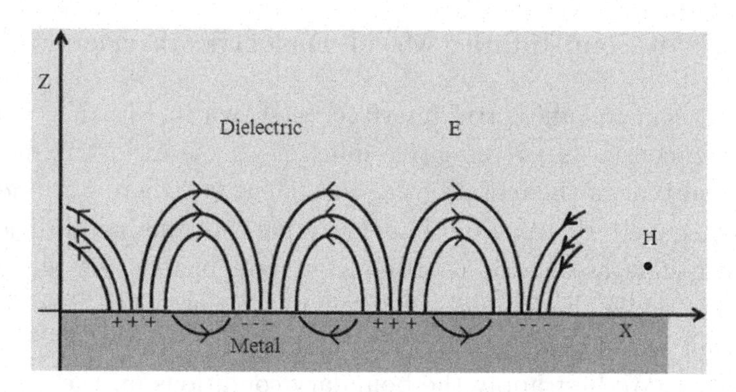

Fig. 2.1. A schematic of the electromagnetic wave and surface charges at a metal–dielectric interface with semi-infinitely extended media on either side.

Eqs. (2.5) and (2.6) as:

$$\frac{d^2H_y}{dz^2} + (k_o^2n^2 - \beta^2)H_y = 0, \tag{2.5}$$

$$\frac{d^2E_y}{dz^2} + (k_o^2n^2 - \beta^2)E_y = 0, \tag{2.6}$$

where k_o $(= 2\pi/\lambda_0)$ is the propagation constant of the incident electromagnetic wave in free space and β is that of the propagating wave in the medium. Since we are interested in the propagating wave solutions confined to the interface, we focus on the solutions of Eqs. (2.5) and (2.6) for the modes with exponential decay in z-direction. Decaying wave solutions of Eqs. (2.5) and (2.6) can be given as:

For $z > 0$,

$$H_y = Be^{-\sqrt{\beta^2 - k_o^2\varepsilon_s}z}, \tag{2.7}$$

$$E_y = Ae^{-\sqrt{\beta^2 - k_o^2\varepsilon_s}z}, \tag{2.8}$$

For $z < 0$,

$$H_y = De^{\sqrt{\beta^2 - k_o^2\varepsilon_m}z}, \tag{2.9}$$

$$E_y = Ce^{\sqrt{\beta^2 - k_o^2\varepsilon_m}z}. \tag{2.10}$$

2.2.1 *Non-existence of SPs for TE modes*

The two boundary conditions for TE waves that are to be applied are given by[4]:

(i) $E_y|_{boundary} = continuous$ and

(ii) $\left.\dfrac{\partial E_y}{\partial z}\right|_{boundary} = continuous.$

Applying the boundary condition (i) to Eqs. (2.8) and (2.10), we obtain:

$$A = C. \tag{2.11}$$

Applying the boundary condition (ii) to the same equations and making use of Eq. (2.11), we obtain:

$$-A\sqrt{\beta^2 - k_o^2\varepsilon_s} = A\sqrt{\beta^2 - k_o^2\varepsilon_m}. \tag{2.12}$$

Squaring both the sides and rearranging the terms, we get:

$$A^2k_o^2(\varepsilon_s - \varepsilon_m) = 0. \tag{2.13}$$

Since ε_s and ε_m are of opposite signs, the term in brackets cannot be zero. This implies that $A = 0$. Thus no surface modes exist for TE polarization.

2.2.2 *Existence of SPs for TM modes*

The two boundary conditions for TM waves that are to be applied are given by[4]

(i) $H_y|_{boundary} = continuous$, and

(ii) $\left.\dfrac{1}{\varepsilon}\dfrac{\partial H_y}{\partial z}\right|_{boundary} = continuous$, i.e.,

$$\left.\frac{1}{\varepsilon_s}\frac{\partial H_y}{\partial z}\right|_{\text{dielectric}} = \left.\frac{1}{\varepsilon_m}\frac{\partial H_y}{\partial z}\right|_{\text{metal}}.$$

Applying the boundary condition (i) to Eqs. (2.7) and (2.9), we get:

$$B = D. \tag{2.14}$$

Applying the boundary condition (ii) to the same equations and using Eq. (2.14) we obtain:

$$\frac{-\sqrt{\beta - k_o^2\varepsilon_s}}{\varepsilon_s} = \frac{\sqrt{\beta - k_o^2\varepsilon_m}}{\varepsilon_m}, \tag{2.15}$$

$$\frac{\beta - k_o^2\varepsilon_s}{\varepsilon_s^2} = \frac{\beta - k_o^2\varepsilon_m}{\varepsilon_m^2}, \tag{2.16}$$

$$\beta^2(\varepsilon_s^2 - \varepsilon_m^2) = k_o^2\varepsilon_s\varepsilon_m(\varepsilon_s - \varepsilon_m), \tag{2.17}$$

$$\beta^2 = \frac{k_o^2\varepsilon_s\varepsilon_m(\varepsilon_s - \varepsilon_m)}{(\varepsilon_s - \varepsilon_m)(\varepsilon_s + \varepsilon_m)}, \tag{2.18}$$

$$\beta = k_o\sqrt{\frac{\varepsilon_s\varepsilon_m}{(\varepsilon_s + \varepsilon_m)}}. \tag{2.19}$$

Equation (2.19) represents the wave-vector of SPs supported by the metal–dielectric interface. Thus:

$$k_{sp} = \frac{\omega}{c}\sqrt{\frac{\varepsilon_s\varepsilon_m}{(\varepsilon_s + \varepsilon_m)}}, \tag{2.20}$$

where ω is the angular frequency and c is the speed of the light in vacuum. The excitation of SPs by TM polarized light is symbolically shown in Fig. 2.1.[5] The charge distributions and field profiles, also shown in Fig. 2.1, point towards the following two important properties of SPs:

(I) The SP mode is coupled to both the dielectric and the metal as the field of the SPs is in both the media.

(II) The field in the metal decays more rapidly than in the dielectric medium, as is symbolically shown by a number of electric field lines. A more accurate and quantitative variation of the field profile is presented later in Fig. 2.3.

The dispersion relation of SPs given by Eq. (2.20) has been plotted in Fig. 2.2 for a gold–air interface. The gray line marked with (i) in Fig. 2.2 is called light line and corresponds to the dispersion of air as $\omega = ck$, while the black curve represents the dispersion curve of the SPs. It shows two branches (ii) and (iii) respectively corresponding

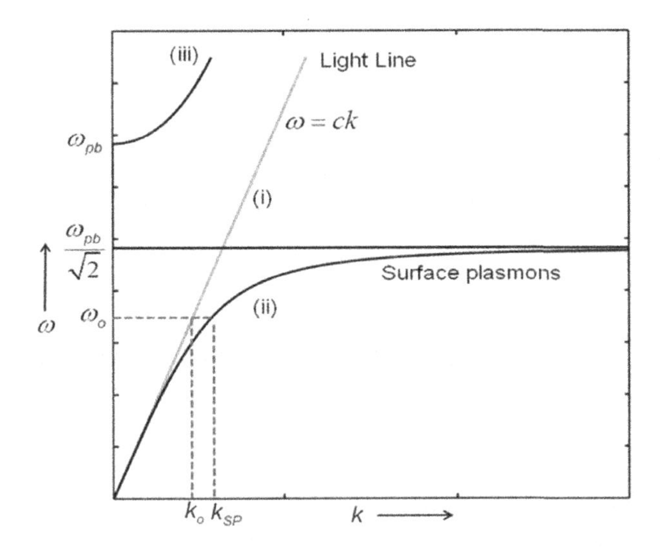

Fig. 2.2. Dispersion of SPs.

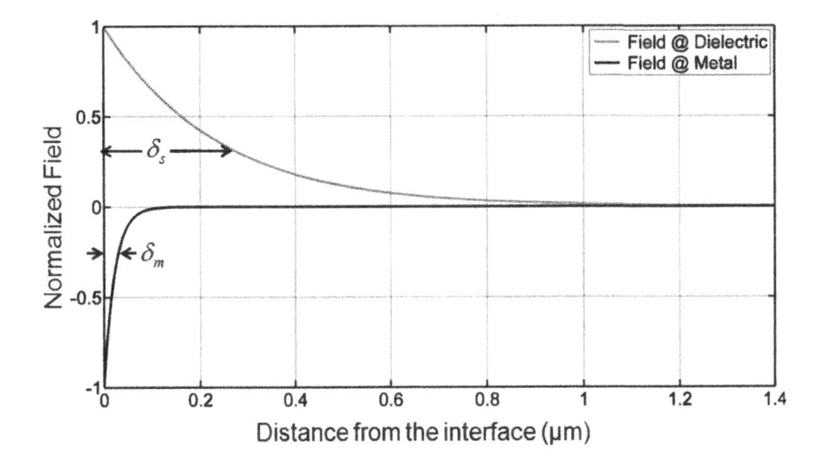

Fig. 2.3. Field profile at a metal–dielectric interface.

to radiative and non-radiative SPs which are bound to the interface. At small wave-vectors, the SP curve approaches the light line, but is still larger than that. So they cannot transform into light and are hence called non-radiative. The most efficient excitation of SPs (or complete transfer of energy) occurs at the higher values of the wave-vector. At the resonance, from Eq. (2.20):

$$-\varepsilon_m = \varepsilon_s.$$

Substituting the value of ε_m from the Drude model (see Appendix A), for negligibly small values of the imaginary part, we get:

$$-1 + \frac{\omega_{pb}^2}{\omega_{SP}^2} = \varepsilon_s,$$

where ω_{pb} is the bulk plasma frequency of the metal and ω_{SP} is the SP resonance frequency. After rearranging the terms, we get:

$$\omega_{SP} = \frac{\omega_{pb}}{\sqrt{1+\varepsilon_s}},$$

For air as a surrounding dielectric medium,

$$\omega_{SP} \approx \frac{\omega_{pb}}{\sqrt{2}}$$

Hence, the asymptotic values of the surface plasma frequencies approach to $\frac{\omega_{pb}}{\sqrt{2}}$, as shown in Fig. 2.2. The volume plasma frequencies for gold (Au) and silver (Ag) are around 8.54 and 8.81 eVs respectively [SOPRA database], while the optical frequencies correspond to 1.55–4.0 eV. However, the plasma frequencies for SPs are a factor of $1/\sqrt{2}$ less than that for volume plasmons. Therefore, if the wave-vectors of the SP wave and that of light become equal, the SPs can be excited. It can be observed from Fig. 2.2 that for a light of frequency ω_0, the wave-vector of SPs is greater than that for light in air. To excite the SPs the wave-vectors of excitation light and the SPs should be equal. Therefore, the light directly incident from the air at the metal–air interface cannot excite SPs. We shall discuss this more rigorously, while introducing the methods for excitation of SPs in Sec. 2.3. Radiative plasmons exist only for the frequencies greater than the plasma frequency, ω_{pb}, and the metal becomes transparent to radiation at such frequencies.

2.2.3 *Field*

The field associated with the SP wave can be written as:

$$E = E_0 \exp[+i(k_x x \pm k_z z - \omega t)], \tag{2.21}$$

where + and − signs are for $z \geq 0$ and $z \leq 0$, respectively and the propagation constant k_z is imaginary which signifies that the field decays exponentially. The wave-vector k_x lies parallel to the x-axis and is given by $k_x = \beta$. In Fig. 2.3 we have plotted field profiles at a metal–dielectric (gold–air) interface.

The field amplitude decays exponentially in both the metal and the dielectric medium. The field amplitude decay in metal is faster than that in the dielectric medium. This can be attributed to the imaginary part of the metal dielectric function which introduces losses.

The two other important parameters associated with SP wave are its penetration depth and propagation length. These are discussed in Secs. 2.2.4 and 2.2.5.

2.2.4 *Penetration depth*

The distance in the material from the interface over which the field amplitude falls to 1/e of its value at the interface is called the penetration depth (δ) of the SP wave.[6] Assuming $|\varepsilon_{m1}| \gg |\varepsilon_{m2}|$, the penetration depths into dielectric (δ_s) and into metal (δ_m) can be written as:

$$\delta_s = \frac{\lambda_0}{2\pi}\left[\frac{\varepsilon_{m1}+\varepsilon_s}{\varepsilon_s^2}\right]^{\frac{1}{2}}, \tag{2.22}$$

and

$$\delta_m = \frac{\lambda_0}{2\pi}\left[\frac{\varepsilon_{m1}+\varepsilon_s}{\varepsilon_{m1}^2}\right]^{\frac{1}{2}}. \tag{2.23}$$

It is quite evident from Fig. 2.3 that the penetration depth in dielectric medium is much larger than that in metal. The values of penetration depth for SP calculated for gold–air and silver–air interfaces for two wavelengths 543 nm and 633 nm of He–Ne laser are tabulated in Table 2.1.

The penetration depth in the dielectric gives us a measure of the length over which SP is sensitive to the changes in the refractive index of the dielectric medium, while that into metal tells about the thickness of the metal film required for the coupling of light incident from the other interface of the metal film.

Table 2.1. Values of penetration depth for $\lambda = 543$ nm and 633 nm at gold/silver–air interface.[7]

	Penetration Depth (nm)			
	$\lambda = 543$ nm		$\lambda = 633$ nm	
	Metal	Dielectric	Metal	Dielectric
Gold	26.67	250.12	26.76	350.18
Silver	23.08	298.52	23.12	414.48

2.2.5 *Propagation length*

The length over which the intensity of the SP wave decreases to 1/e of its maximum value is called the propagation length (L_{sp}).[6] The dielectric constant of the metal, ε_m, being a complex number, the propagation constant of the SP wave, k_{sp}, will also be a complex number having real and imaginary parts k'_x and k''_x i.e.:

$$k_{sp} = k_x = k'_x + ik''_x, \tag{2.24}$$

where:

$$k'_x = \frac{\omega}{c}\left(\frac{\varepsilon_{m1}\varepsilon_s}{\varepsilon_{m1} + \varepsilon_s}\right)^{1/2}, \tag{2.25}$$

and

$$k''_x = \frac{\omega}{c}\left(\frac{\varepsilon_{m1}\varepsilon_s}{\varepsilon_{m1} + \varepsilon_s}\right)^{1/2} \frac{\varepsilon_{m2}}{2(\varepsilon_{m1})^2}, \tag{2.26}$$

Therefore, the intensity of the SP wave propagating along the interface decreases as $e^{-2k''_x x}$. Thus the propagation length is given by[6]:

$$L_{sp} = (2k''_x)^{-1}. \tag{2.27}$$

The metals having a large (negative) value of the real part of the dielectric constant and a small value of the imaginary part possess larger propagation lengths. The propagation length suggests the limiting size of the photonic/plasmonic structure one can have. The calculated values of propagation length for SPs for gold–air

Table 2.2. Values of propagation length for $\lambda = 543\,\text{nm}$ and 633 nm for gold/silver–air interface.[7]

	$\lambda = 543\,\text{nm}$	$\lambda = 633\,\text{nm}$
	Propagation Length (μm)	
Gold	10.17	15.34
Silver	29.82	43.67

and silver–air interfaces at two wavelengths 543 nm and 633 nm are tabulated in Table 2.2.

2.3 Excitation of SPs

As confirmed by Eqs. (2.13) and (2.19), SPs are TM (or p-) polarized and hence can be excited by TM polarized light. For the excitation of SPs, the wave-vector of the excitation light along the metal–dielectric interface should be equal to that of SPs, a condition called the resonance condition. However, as seen in Fig. 2.2, the dispersion curves for the SPs and the incident light do not cross at any frequency and hence the wave-vectors of the incident light and the SPs cannot be equal. This implies that the SPs cannot be excited by a beam of light directly incident on the interface. To excite the SPs, the wave-vector of the incident light must be increased to recover the mismatch between wave-vector or phase of it and SPs. The most general methods for increasing the wave-vector to recover the phase mismatch utilize prism, waveguide, and grating based configurations. We discuss these methods, in detail, in Secs 2.3.1–2.3.3.

2.3.1 *Prism-based method*

Before discussing this method we shall discuss the reflection and refraction phenomena of a plane wave at an interface. Consider two dielectric media with dielectric constants ε_1 and ε_2 ($\varepsilon_1 > \varepsilon_2$). If a plane wave is incident at the interface from the denser medium (ε_1) at an angle less than the critical angle

$[\theta_c = \sin^{-1}(\sqrt{\varepsilon_2}/\sqrt{\varepsilon_1})]$ then both reflection and refraction take place as shown in Fig. 2.4(a). For angles greater than the critical angle, the plane wave is totally reflected in the denser medium and no light is transmitted in the rarer medium. A plot of normalized intensity of the reflected light as a function of angle of incidence is shown in Fig. 2.4(b).

According to Fig. 2.4(b) there is no electromagnetic field in rarer medium under the total internal reflection condition. In reality this is not the case. There exists an electromagnetic field in the rarer medium called evanescent field. Its amplitude is the maximum at the interface and decreases exponentially in the rarer medium. This field is associated with a special type of electromagnetic wave called evanescent wave which propagates along the interface. It is derived from the Maxwell's equations and by applying boundary conditions at the interface. Let us go back to Fig. 2.4(b). The electric field of the incident light beam lying in the plane of incidence and incident on the interface can be written as:

$$\mathbf{E}_1 = \mathbf{E}_{1o} e^{i(k_{1x}x + k_{1z}z - \omega t)}, \tag{2.28}$$

This is the equation of a plane wave with ω as the angular frequency and the x and z components of propagation constant are k_{1x} and k_{1z} respectively. These propagation constants satisfy the following relation:

$$k_1^2 = k_{1x}^2 + k_{1z}^2 = \left(\frac{\omega}{c}\right)^2 \varepsilon_1, \tag{2.29}$$

where c is the speed of light in vacuum.

For angle of incidence θ, k_{1x} can be written as:

$$k_{1x} = \left(\frac{\omega}{c}\right) \sqrt{\varepsilon_1} \sin\theta, \tag{2.30}$$

The electric field in the rarer medium can be written as:

$$\mathbf{E}_2 = \mathbf{E}_{2o} e^{i(k_{2x}x + k_{2z}z - \omega t)}, \tag{2.31}$$

with

$$k_2^2 = k_{2x}^2 + k_{2z}^2 = \left(\frac{\omega}{c}\right)^2 \varepsilon_2, \tag{2.32}$$

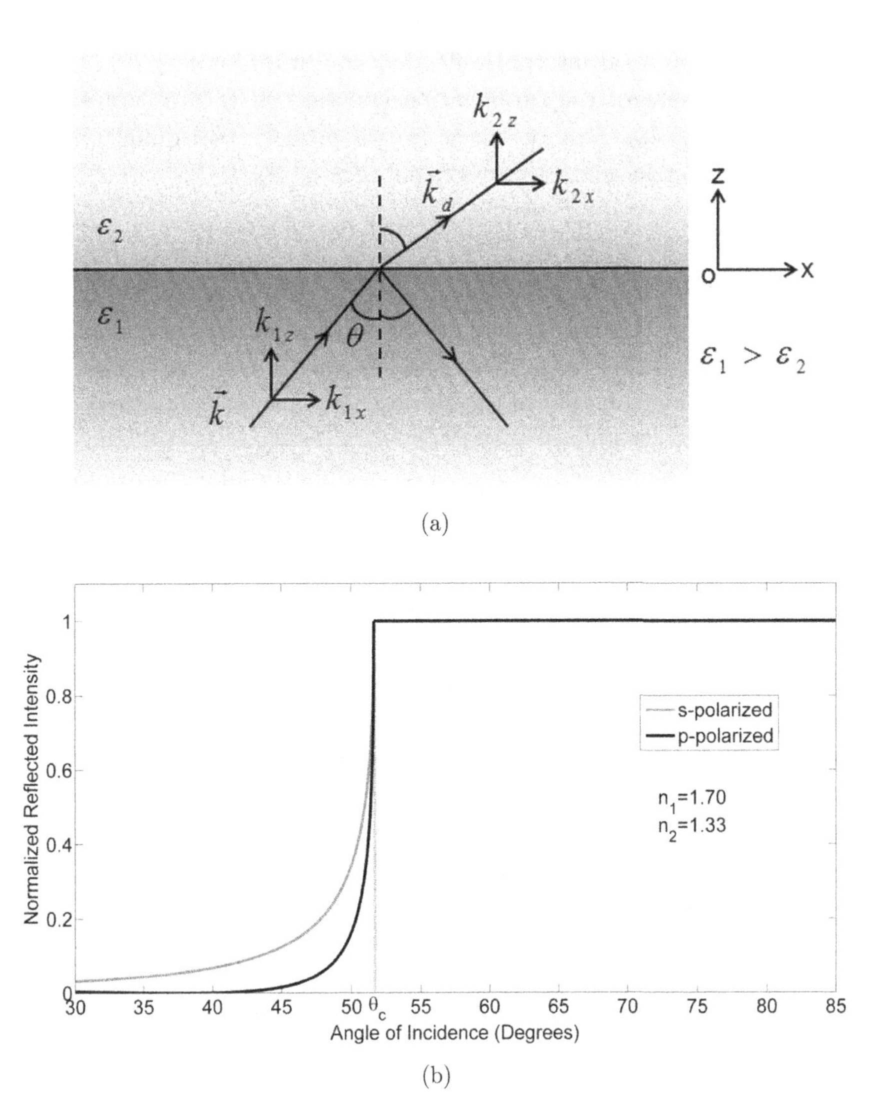

Fig. 2.4. (a) Reflection and refraction of a plane wave at an interface of a denser–rarer media. (b) Normalized intensity of reflected light as a function of angle of incidence at the interface.

The boundary condition at the interface of two media requires $k_{1x} = k_{2x}$. Thus:

$$k_{2z} = \pm\frac{\omega}{c}(\varepsilon_2 - \varepsilon_1 \sin^2\theta)^{1/2}. \tag{2.33}$$

For $\theta > \theta_c$, the total internal reflection condition, and $\varepsilon_2 < \varepsilon_1 \sin^2 \theta$ the z-component of the propagation constant in the rarer medium takes the form $k_{2z} = \pm i\alpha$, where α is a positive real quantity. For $z > 0$, the electric field can be written as:

$$\mathbf{E}_2 = \mathbf{E}_{2o} e^{i(k_{2x}x - \omega t)} . e^{-\alpha z} \tag{2.34}$$

This is the field equation of an evanescent wave that propagates along the x-direction (interface) and its amplitude decays exponentially along the z-direction (away from the interface) in the rarer medium. The propagation constant of the evanescent wave, in the present case, is given by Eq. (2.30).

Now we will go back to the prism-based method for the excitation of SPs. In this method, initially two configurations were proposed. These are called Otto and Kretschmann configurations.[8,9] In both the configurations, a high refractive index prism is used. The p-polarized light from a monochromatic source is incident at the base of the prism at an angle greater than the critical angle through one of the faces and is collected at the other face of the prism which is further fed into an optical power meter. The incidence at an angle greater than the critical angle results in the generation of an evanescent wave which propagates along the interface of the prism base and the medium in contact. The evanescent wave is used to excite SPs in both kinds of configurations. In Otto configuration, an infinitesimal (~nm size) gap between a metal film and the prism base is maintained and the dielectric or sensing medium of refractive index lower than the refractive index of the material of the prism is filled in the gap as shown in Fig. 2.5(a). The evanescent field generated at the prism–dielectric medium interface excites SPs at the metal–dielectric medium interface. The resonance or the complete transfer of light to SPs occurs when the propagation constants of the evanescent wave and the SP wave become equal. This occurs at a particular angle of incidence, greater than the critical angle, and in this condition a drop in the reflected power is observed. Thus, we get a minimum in the reflected power versus angle of incidence plot. The angle corresponding to minimum reflected power is called the resonance angle. Such a configuration is not very much in use because

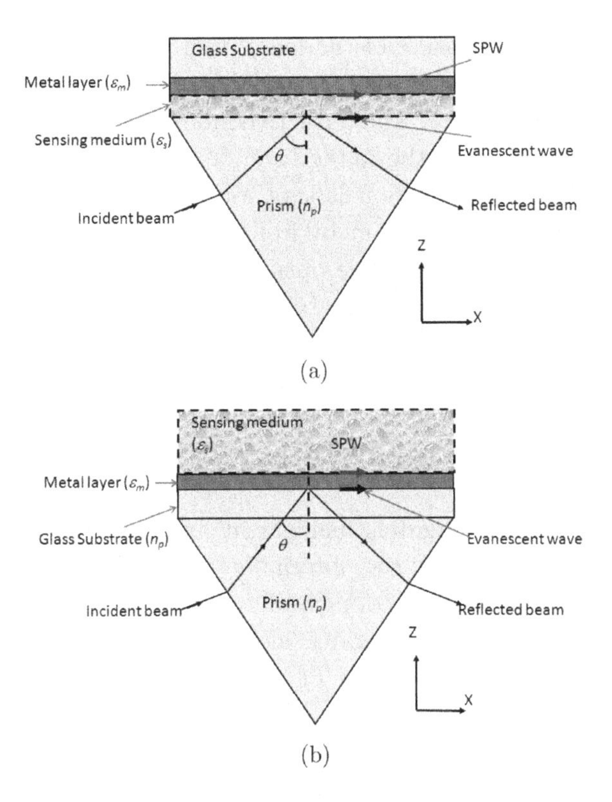

Fig. 2.5. (a) Otto. (b) Kretschmann configuration.

of the practical difficulty of maintaining the nano-size gap between the prism base and the metal surface. The Otto configuration is generally used for calculating the surface roughness of metal films. In another configuration called Kretschmann configuration, the base of the prism is coated with a thin film of the metal and the dielectric medium is kept in contact of the other side of the metal film. The evanescent wave propagating along the interface of prism and metal layer excites the SPs at the metal–dielectric interface. The complete excitation occurs when the wave-vectors of the two waves match. In practice, a glass slide of the refractive index similar to that of the prism having thin metallic coating on one side and refractive index equal to that of the prism is kept in contact of the base of the prism with the help of an index matching liquid as shown in Fig. 2.5(b).

Doing so eases the coating of the metal film and its integration into the optical setup, as one does not need to replace the prism from the optical setup for coating. The refractive index of the substrate/prism poses the upper limit of the refractive index of the dielectric medium that can be used and hence it is kept as large as possible to achieve larger range for sensing. The metal film coated substrate or glass slide is used to avoid disturbing the optical setup for different coatings. As mentioned earlier the collimated p-polarized light is incident on the prism–metal interface through one of the faces of the prism and the reflected light is collected from the other face of the prism. The Kretschmann configuration is used mostly due to its practical easiness. Next, we discuss the Kretschmann configuration in more detail.

When p-polarized light is incident at an angle greater than the critical angle, it gets totally internally reflected from the surface of the glass substrate in contact of metal film and generates an evanescent wave that propagates along the substrate–metal layer interface. When the thickness of the metal film is of the order of tens of nanometer the field of this evanescent wave overlaps to the metal-sensing medium interface. The propagation constant of the evanescent wave, in terms of prism parameters, can be written as[1]:

$$k_{ev} = \frac{\omega}{c}\sqrt{\varepsilon_p}\sin\theta = k_o n_p \sin\theta, \tag{2.35}$$

where ε_p and n_p are respectively the dielectric constant and the refractive index of the material of the prism and θ is the angle of incidence of the light beam. It may be understood that the wave-vector of the light reaching the interface gets increased, while travelling in the prism due to its high refractive index. The resonance condition, for the excitation of the SPs at the metal–dielectric (sensing medium) interface, can then be written as:

$$\sqrt{\varepsilon_p}\sin\theta = \sqrt{\frac{\varepsilon_{m1}\varepsilon_s}{\varepsilon_{m1}+\varepsilon_s}} \tag{2.36}$$

In the Eq. (2.36) $\varepsilon_p > \varepsilon_s$ and ε_{m1} is negative, hence for some value of angle greater than the critical angle, Eq. (2.36) or the resonance

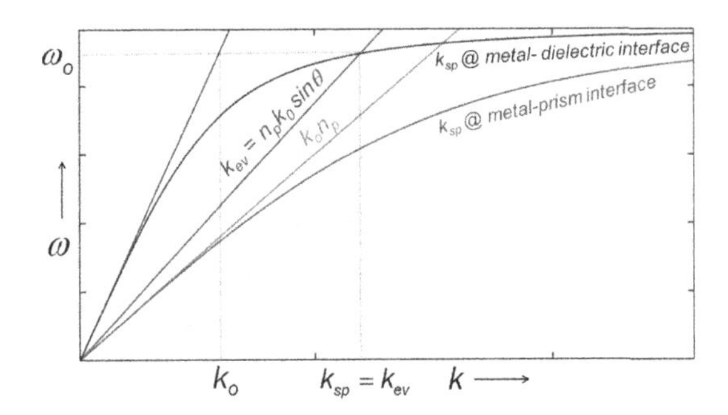

Fig. 2.6. Dispersion curves for a Kretschmann configuration.

condition will be satisfied. In that case, the dispersion curves of SPs and evanescent wave will cross each other as shown in Fig. 2.6.

It shall also be noted that the prism–metal interface also supports SP modes. These modes, however, cannot be excited because the wave-vector of SPs at this interface is greater than that for the incident light, as can easily be observed from Fig. 2.6. A change in the refractive index of the dielectric medium around the metal film leads to the change in the resonance parameter, for example, the angle of incidence; a property used for sensing applications.

2.3.2 *Waveguide-based method*

In the previous method the excitation of the SPs is based on total internal reflection at the prism base. Therefore, the prism can be replaced by a waveguide layer or core of an optical fiber because the light guidance in waveguide or fiber is due to the total internal reflection. Thus the method of excitation of SPs using waveguide is the same as that of the prism-based Kretschmann configuration. In the case of waveguide, a small length of the wave-guiding layer in the middle is coated with a metal layer. The evanescent field of the guided modes of the light propagating inside the waveguide excites SPs at the metal–dielectric interface. A schematic of the excitation of the SPs in waveguide geometry is shown in Fig. 2.7. The maximum refractive index of the dielectric sensing medium is limited to that of

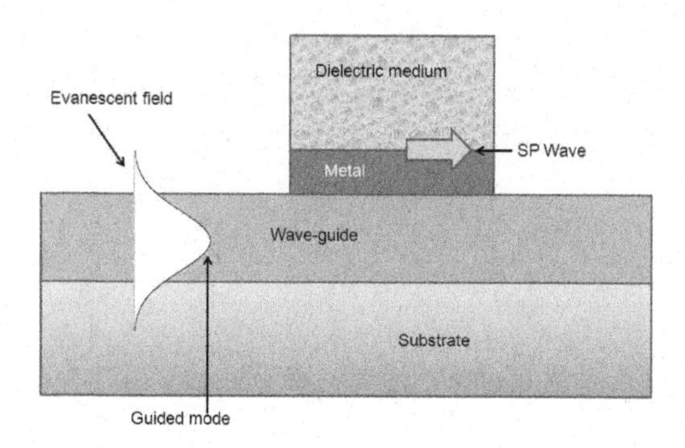

Fig. 2.7. Excitation of SPs using optical waveguide method.

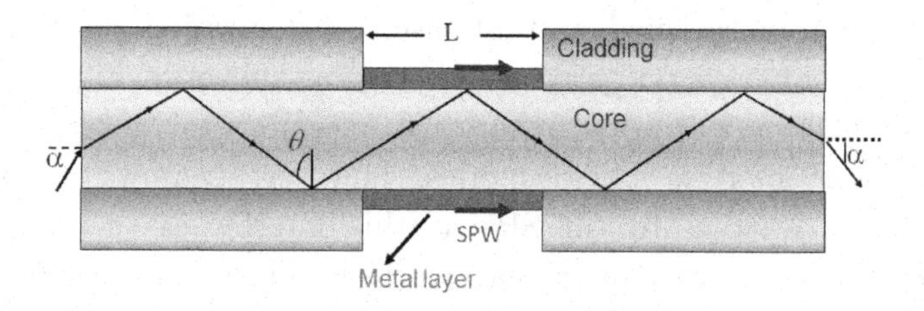

Fig. 2.8. Excitation of SPs using optical fiber.

the waveguide layer; posed by the Kretschmann geometry. Optical fiber can also replace 2-dimensional optical waveguide as shown in Fig. 2.8. In the case of optical fiber a small portion of the cladding is removed from the middle and the unclad core is coated with a thin layer of metal. Since in the fiber a range of guided rays propagates the resonance condition of equality of wave-vectors is achieved by varying the wavelength of the coupling light in the fiber. In other words, a polychromatic light is guided through the core of the fiber which excites SPs and the spectrum of the light transmitted off the output end of the fiber has a dip at a particular wavelength called resonance wavelength. At resonance wavelength, the wave-vectors of the evanescent wave propagating along the core-metal interface match with the wave-vector of the SP wave propagating along the metal–dielectric sensing medium.

As mentioned earlier, TM or p-polarized light is used to excite SPs but in the case of an optical fiber it is difficult to maintain the polarization of the light launched in the fiber due to effects like cylindrical geometry, fiber bending, etc. Thus, in the case of an optical fiber an unpolarized light is launched and hence both TE and TM polarized light propagate in the fiber. Since the TM-polarized light excites the SPs, no effect will be on the TE polarized light. Thus, the net transmitted light received at the other end of the fiber will be the sum of the unaffected TE and SP affected TM-polarized lights. Because of this the minimum normalized transmitted power corresponding to the resonance wavelength at the output end of the fiber cannot be less than 0.5.

In the case of optical fiber, the excitation of SPs by evanescent wave depends on various parameters such as wavelength of light, fiber parameters like core diameter, doping material, numerical aperture, fiber geometry, metal and its thickness used. The mechanism of coupling of evanescent field with SPs is different for single mode and multi-mode fibers. Further, the coupling also depends on the penetration depths of evanescent field which again depends on the numerical aperture of the fiber and the fiber geometry. For example, the penetration depth increases if the fiber is bent or tapered. The effects of these kinds of parameters on the SP resonance will be discussed in more detail in later chapters.

2.3.3 *Grating-based method*

In general gratings are used to find the spectra of light sources; however, they can also be used for the excitation of SP. Consider a 1-dimensional periodic structure written on the metal surface as shown in Fig. 2.9. A monochromatic p-polarized light is incident through the dielectric medium on the grating surface at an angle θ. The grating diffracts light in different directions identified with orders. In addition to diffraction, light is also reflected specularly from the surface. If the intensity of this specularly reflected light is measured as a function of angle of incidence, a dip in the intensity can be obtained if the grating element is suitably selected. This occurs due to the excitation of SPs at the dielectric-grating interface. For such a configuration, the dielectric medium must be transparent. In

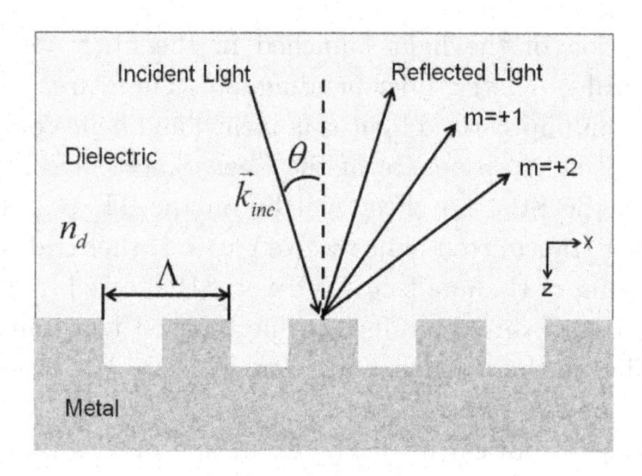

Fig. 2.9. Excitation of SPs using 1-dimensional metallic grating.

a grating-based SPR setup, the wave-vector of the incident light is increased by the higher order diffraction modes. The mode with a wave-vector equal to that of SPs is found absent in the collected light resulting in the dip. The component of the wave-vector of the diffracted waves along the interface is given by:

$$k_x = (k_{\text{inc}})_x \pm \Delta k_x, \tag{2.37}$$

or, more explicitly:

$$k_x = (k_{\text{inc}})_x \pm m\left(\frac{2\pi}{\Lambda}\right) = \frac{2\pi}{\lambda} n_d \sin\theta \pm m\left(\frac{2\pi}{\Lambda}\right), \tag{2.38}$$

where m is the order of diffraction, can have both positive and negative integer values, and Λ is the grating element. For positive values of m, the resonance condition can be written as:

$$k_{SP} = \frac{2\pi}{\lambda} n_d \sin\theta + m\left(\frac{2\pi}{\Lambda}\right). \tag{2.39}$$

The earlier condition can be satisfied only for a particular value of the angle of incidence. Thus the role of the grating is to supply a wave-vector of $m(\frac{2\pi}{\Lambda})$ to $k_x = (k_{\text{inc}})_x$ so that it can match with the k_{SP}.

2.4 SP Modes of a Thin Metal Film

We have seen above that in Kretschmann configuration, two SP modes are available; one at the prism–metal interface and the other at the metal–dielectric medium interface. This is a case of SP modes of a thin metal film sandwiched between two dielectric media of different refractive indices propagating at upper and lower interfaces. For a case when the refractive indices of the dielectric media on both the sides of the thin metal film are almost the same, we get two SP modes corresponding to each interface with almost similar propagation constants. If the metal film is thick enough then the two SP modes are independent of each other. However, when the metal film is thin, then there occurs a metal-thickness dependent coupling between these two SP modes, which results in two different modes called symmetric and anti-symmetric SP modes. The electric field distribution in the case of symmetric and anti-symmetric SP modes are schematically shown in Fig. 2.10. The symmetric mode is the result of superposition of two modes of two interfaces when these are in phase, while anti-symmetric mode is due to the superposition of two modes when these are out of phase. The propagation length of the symmetric mode is large and is called long-range SP mode, while the anti-symmetric mode is called short range surface plasomon (SRSP) due to short propagation length. The propagation length depends on

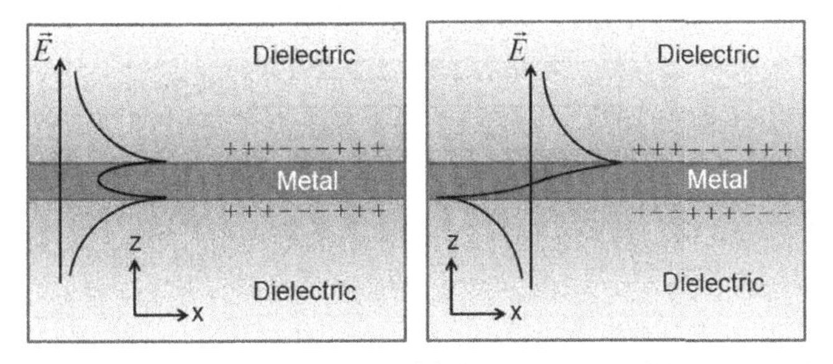

(a) Symmetric Plasmon Mode (b) Anti-symmetric Plasmon Mode

Fig. 2.10. (a) Symmetric. (b) anti-symmetric SP modes of a dielectric–metal–dielectric structure.

the metal and the dielectric used, thickness of the metal film, and the wavelength of the light.

A rigorous mathematical formulation of such a system is out of scope of the present book and can be seen elsewhere.[10,11] We restrict ourselves here to the discussion of LR- and SR-SPs occurring as a result of such coupling.

2.5 Long and Short Range Surface Plasmons

In the previous section we have defined LR- and SR-SPs. To excite these, a thin dielectric layer (~500 nm) of almost similar refractive index as that of the dielectric sensing medium is coated on the glass substrate before coating of the metal film (Fig. 2.11). Such a dielectric–metal–dielectric structure supports two super modes discussed earlier: symmetric and anti-symmetric. The symmetric modes have a smaller propagation constant and anti-symmetric ones have larger. As mentioned in Sec. 2.4, the anti-symmetric modes are called short range surface plasmons (SRSPs) while the symmetric modes are called long range surface plasmons (LRSPs).

Figure 2.12 shows the SPR curves for a LRSPR-based Kretschman configuration (SF11/Teflon/silver/1.33–1.34) at 653 nm wavelength and 500 nm thickness of Teflon layer. The refractive indices of Teflon layer and dielectric sensing medium are nearly the same. In the

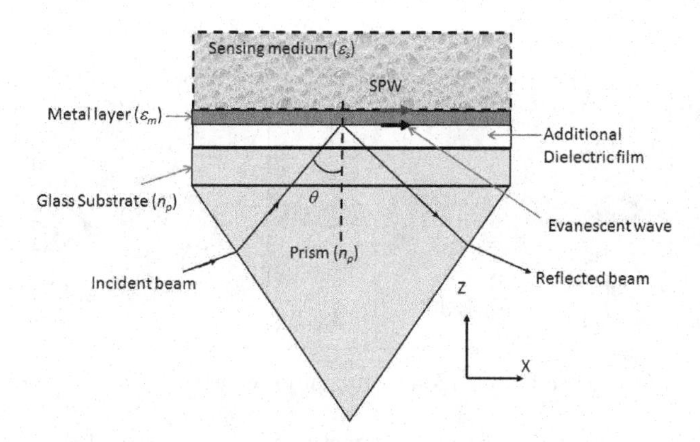

Fig. 2.11. Kretschmann configuration for long range surface pasmon resonance (LRSPR).

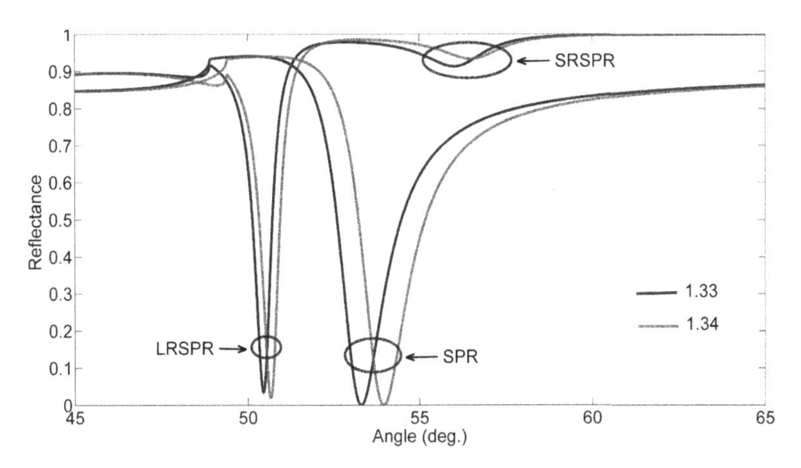

Fig. 2.12. LRSP and SRSPs @653 nm for (SF11/Teflon-500 nm/Ag-46.6 nm/Analyte-1.33:1.34) configuration. For conventional SPR, Teflon thickness = 0 nm.

same figure a conventional SPR curve has also been plotted which corresponds to zero thickness of Teflon layer. As expected, two SPR minima are observed when a Teflon layer is introduced in the configuration; one corresponding to long range surface plasmon resonance (LRSPR) and the other to short range surface plasmon resonance (SRSPR). In the case of the conventional SPR curve one observes only one minimum. From the same figure, it can also be predicted that the width of the LRSP curve is very sharp implying higher detection accuracy of the resonance angle in comparison to conventional SPs and SRSPs. However, the shift in resonance angle in the case of LRSPR is slightly smaller than that for SPR and SRSPR. But due to the low propagation loss, LRSPs are used as an alternative to conventional SPs for sensing applications.

2.6 Nearly Guided Wave SPR (NGWSPR)

Such a configuration was recently coined by Shalabney and Abdulhalim.[12] In such a configuration, a very thin layer (~10 nm) of a high index dielectric film (Si or ZrO_2) is coated on top of the metal layer (Fig. 2.13). We present next in Fig. 2.14 our simulated results of electric field profile for a NGWSPR configuration to present a

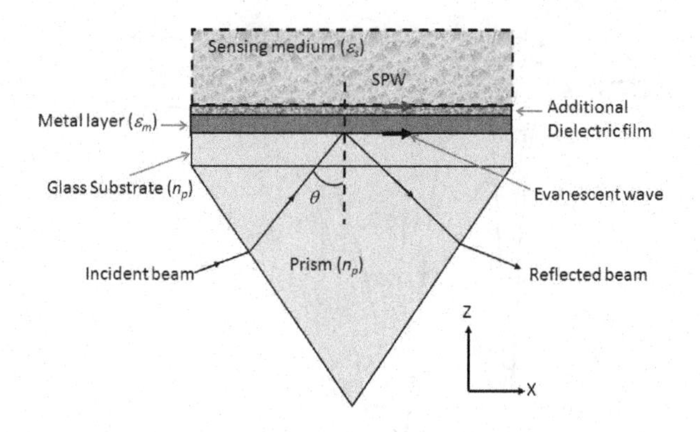

Fig. 2.13. NGWSPR — in Kretschmann configuration.

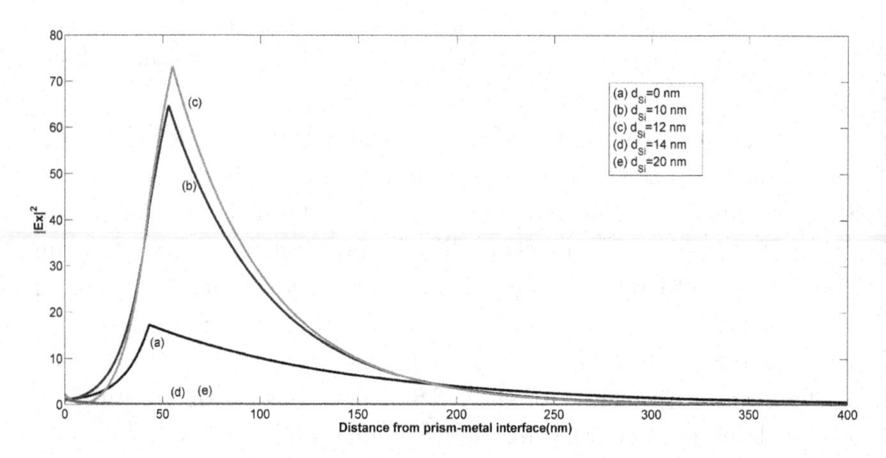

Fig. 2.14. Field profiles for a NGWSPR structure as compared to conventional SPR configuration. The simulations were carried out for (SF11/Ag-43 nm/Si-0:20 nm/water) configuration at 653 nm.

comparison between the field profiles of NGWSPR with conventional SPR. The simulations have been carried out for (SF11/Ag/Si/water) configuration at 653 nm wavelength of the light. The thickness of the silver layer is 43 nm, while the thickness of the silicon layer is varied from 0 to 20 nm. It is observed that the field amplitudes get highly enhanced (Fig. 2.14) at the interface of the outer layer of silicon and the dielectric sensing medium. It is quite evident from Fig. 2.14 that the maximum enhancement occurs only for an

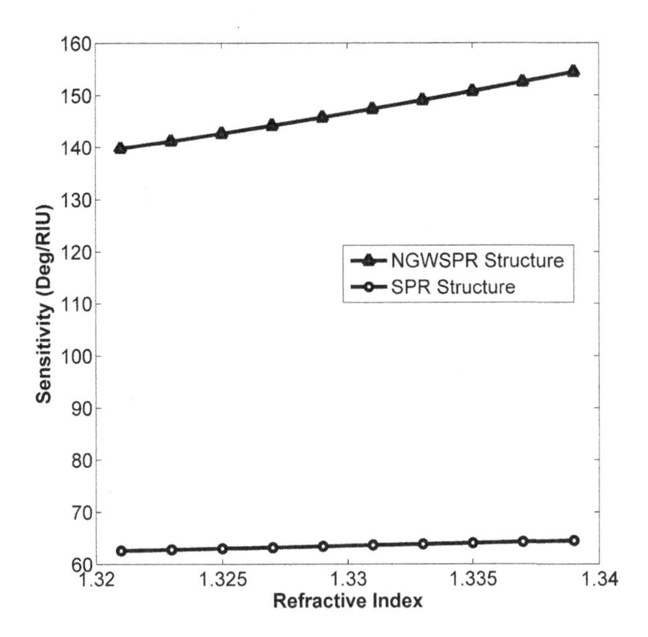

Fig. 2.15. Sensitivity comparison of conventional and NGW–SPR sensors. For NGWSPR, (SF11/Ag-43 nm/Si-10 nm) structure was chosen and for conventional SPR, $d_{Si} = 0$.

optimum value of the Si cover thickness. The field enhancement leads to an increase in the coupling between the SP mode and the analyte, which leads to an increase in the sensitivity. In Fig. 2.15, we have plotted the variation of sensitivity with refractive index for both the SPR and NGWSPR configuration. In spectral mode of operation, overall Figure of merit (FOM) is very high as compared to conventional SPR sensors. Such structures have been prepared on fiber optic configuration and rigorously studied both theoretically and experimentally for biosensing applications which we shall discuss in more detail in chapters 6 and 7.

2.7 Interrogation Techniques

For the excitation of SP, its wave-vector should match with the wave-vector of the excitation light. Any change in the refractive index of the sensing medium leads to a change in the wave-vector of SP and hence the required wave-vector of the excitation light. To find the change in the dielectric one has to find the change taken place in the wave-vector of the excitation light. The wave-vector of the excitation light depends on the angle of the incident light and/or the wavelength of the excitation light. Hence, any change in the wave-vector of SPs results in the change in the resonance parameter (angle or wavelength) of the excitation light. For example, a small change in the dielectric constant or the refractive index of the dielectric sensing medium leads to a change in the wave-vector of SPs which will be reflected by the change in one of the resonance parameters. The change in the dielectric properties of the dielectric sensing medium can be determined by precisely measuring the change in the resonance parameter. This is the basic principle of the SP resonance-based sensors. In addition to angle and wavelength, there are two other characteristic parameters, intensity and phase of the incident light measurement of which can also predict the change in the dielectric properties of the sensing medium. On the basis of change in the light characteristic, the measurement techniques can be classified as angular, wavelength, intensity and phase interrogations. These interrogation methods are discussed as follows, in detail.

2.7.1 *Angular interrogation*

As is evident from Fig. 2.6, the resonance condition given by Eq. (2.36) must be satisfied for the excitation of SPs. For a p-polarized monochromatic collimated light beam (fixed λ and n_p), the resonance condition is satisfied at a particular angle of incidence called resonance angle (θ_{res}). At resonance, the energy of the incident light gets efficiently transferred to the SPs resulting in a sharp decrease in the reflected light intensity. Therefore, in an experiment with monochromatic light, the angle of incidence at the base of the prism is varied and a sharp dip is observed at a particular angle in the reflected light intensity versus angle of incidence plot. The angle of incidence which corresponds to minimum reflectivity is called resonance angle. Thus in a sensing device based on angular interrogation method, the intensity of the reflected light at the output

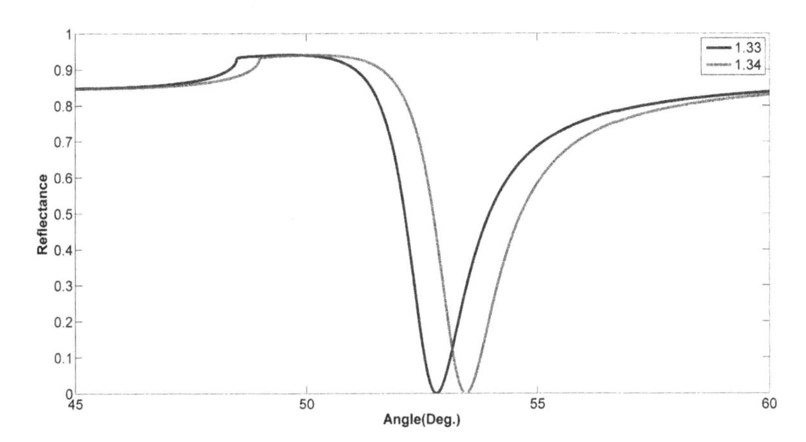

Fig. 2.16. SPR curves for varying refractive indices of the sensing medium. The simulations were carried out for SF11/Ag-47 nm/sensing medium at 653 nm.

end is measured as a function of angle of incidence θ for fixed values of the wavelength of light source, metal layer thickness, and dielectric constant of the sensing medium. In Fig. 2.16 we have plotted the SPR curves in angular interrogation mode for two different refractive indices, 1.33 and 1.34, of the sensing medium. It is observed that the resonance angle moves to higher angles with an increase in the refractive index of the sensing medium. A graph between the resonance angle and the refractive index of the dielectric sensing medium serves as the calibration curve of the sensor. Based on the calibration curve, the resonance angles can be translated to the unknown refractive indices of the test samples.

2.7.2 *Spectral interrogation*

In spectral interrogation mode of operation, the angle of incidence of the light beam is kept constant and is greater than the critical angle while, instead of a monochromatic source, a polychromatic light source is used. In such a scheme, the resonance condition in Eq. (2.36) is satisfied at one particular wavelength of the light source where the maximum transfer of energy from the evanescent wave to the SP wave takes place resulting in a dip in the spectrum of the reflected light beam. The wavelength corresponding to minimum reflected

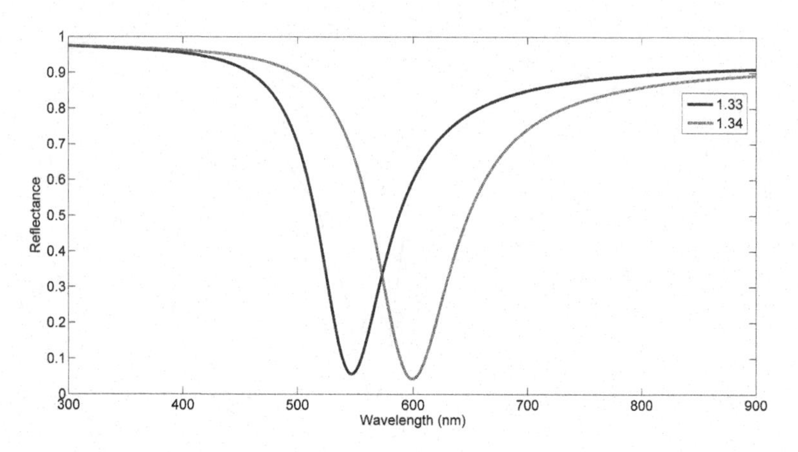

Fig. 2.17. SPR spectra for varying refractive indices of the sensing medium. The simulations were carried out for SF11/Ag-47nm/sensing medium. The dispersion relation for SF11 was taken from refractive index info database and Drude model was used for silver. The angle of incidence was taken as 53.58°.

intensity is called the resonance wavelength (λ_{res}). In Fig. 2.17 we have plotted the SPR spectra for two different refractive indices of the sensing medium. It is observed that when the refractive index of the sensing medium is increased from 1.33 to 1.34, the resonance wavelength shows a red shift from 546 to 599 nm for the particular set of parameters used for the simulation. In the case of sensors utilizing spectral or wavelength interrogation a plot of resonance wavelength versus refractive index of the sensing medium called the calibration curve is used to find the refractive index of the unknown dielectric medium. The advantage of such a scheme is that one does not need to scan over a wide range of angles, and hence no mechanical moving parts are required. But such a setup becomes costlier due to use of spectrometer. The other problem associated with such a setup is the poor collimation of the polychromatic beam, which, eventually, leads to widening of the spectrum, as the resonance condition now gets satisfied over a small range of wavelengths due to a finite and small range of angles of incidence. Sensors utilizing waveguide-based geometry, such as fiber optic SPR sensors, utilize spectral interrogation where all the guided rays are launched and a polychromatic light source is used.

2.7.3 *Intensity interrogation*

In this method, the configuration is similar to what is used in angular interrogation. The difference is that, in this case, the intensity of the reflected light beam is measured as a function of the refractive index of the sensing medium at a fixed value of the angle of incidence greater than the critical angle. In Fig. 2.18, we have plotted simulated SPR curves for different refractive indices of the sensing medium in the Kretschmann configuration similar to Fig. 2.17 for the light of wavelength 653 nm. As mentioned in the angular interrogation method, the resonance angle increases as the refractive index of the sensing medium increases. In the figure, two vertical lines (a) and (b) have been drawn corresponding to two angles of incidence. If the angle of incidence in the setup is kept fixed and the intensity as a function of refractive index of the sensing medium is measured then in the case of angle (a) the intensity increases as the refractive index increases for the refractive index greater than 1.330, while in the of case (b) the intensity decreases as the refractive index increases. Thus this method can also be used for the measurement of the refractive index of the sensing medium. In this interrogation method, it is very important to define the range of operation and the use of one of the two ranges of operation. Random choice of incidence angle leads to change in the calibration curve.

In the intensity interrogation method discussed earlier, a monochromatic source is used and intensity is measured at one particular angle of incidence as a function of the refractive index of the sensing medium. However, if we use wavelength interrogation method and plot SPR curves for different refractive indices of the dielectric sensing medium then similar kinds of curves, plotted in Fig. 2.18, with different resonance wavelengths are obtained. In this situation when angle of incidence is fixed one can draw two vertical lines and can choose two wavelengths similar to angles in Fig. 2.18. For one wavelength, the reflected intensity increases with the refractive index, while for the other wavelength it decreases with the refractive index of the dielectric sensing medium. In some of the cases intensity is measured for two different wavelengths and the ratio

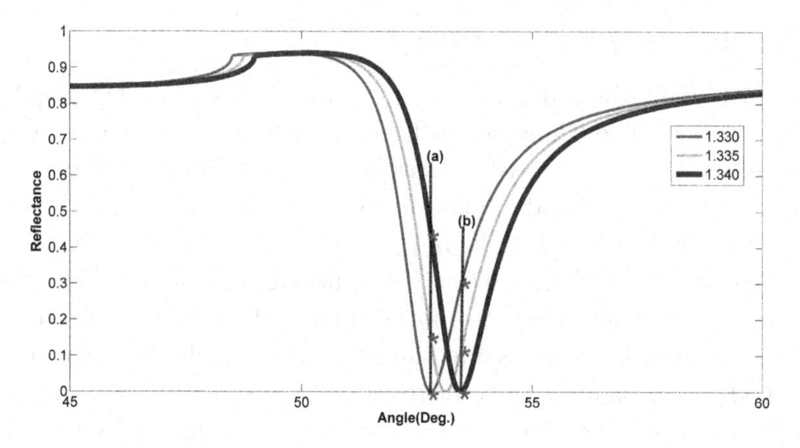

Fig. 2.18. SPR curves at different refractive indices and the scheme of intensity interrogation.

of intensities is determined as a function of refractive index.[13] In this method the fluctuation of intensity due to other reasons get cancelled. For this kind of system, two light emitting diodes (LEDs) of different wavelengths are used and intensities are measured at these two wavelengths. The knowledge of change in differential intensities or the ratio is used to determine the change in the dielectric constant of the medium.

2.7.4 *Phase interrogation*

Let us again consider the prism-based Kretschmann configuration used in angular and intensity interrogation schemes with p-polarized monochromatic light. In phase interrogation scheme, both p- and s-polarized incident lights of fixed wavelength are used and the phase difference between the reflected lights of these two polarizations is measured as a function of angle of incidence for a given refractive index of the sensing medium. Figure 2.19 shows the variation of the difference in phase between p- and s-polarized light with respect to angle of incidence (dark curve). In the background of the figure, we have plotted the SPR curve (faded curve) as well to clearly understand the behavior near the resonance angle. It is observed that the variation of phase difference is very sharp around the

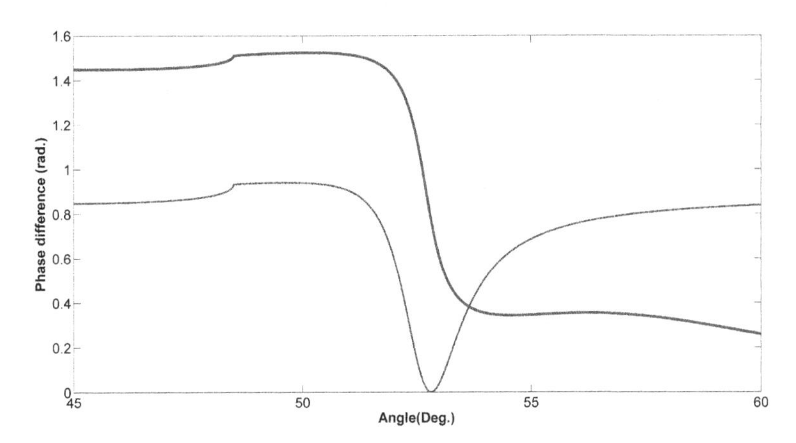

Fig. 2.19. Variation of phase difference with angle of incidence (dark curve). The faded curve corresponds to conventional SPR curve.

resonance angle. Thus the phase interrogation appears to be more sensitive. However, since the phase variation is extracted from the measurements of intensity, small fluctuations in the source power may lead to wrong phase predictions. Hence the credibility of phase measurements has been questioned from time to time.

2.8 SPR Imaging (SPRI)

In a SPRI system, rather than measuring the spectrum or power, the image of the reflected light at the output edge of the prism is recorded by a camera. A schematic of a SPRI system is shown in Fig. 2.20. Although variety of SPRI designs are avialble in the literature and in the market (such as SPREETA), we discuss here a simple diverging beam SPRI setup. In such a setup, light is launched at a range of angles from one of the faces of the prism and the light collected from the other face of the prism is fed to a camera with the help of a convex lens. Light launched at an angle corresponding to the one satisfying the surface plasmon resonance (SPR) condition gets absorbed and we get a dark line corresponding to SPR in the image. A SPR image is shown in Fig. 2.21. The recorded image is then numerically translated into angles to find the corresponding resonance angle. Any change in the refractive index of the sensing medium leads to shift in the

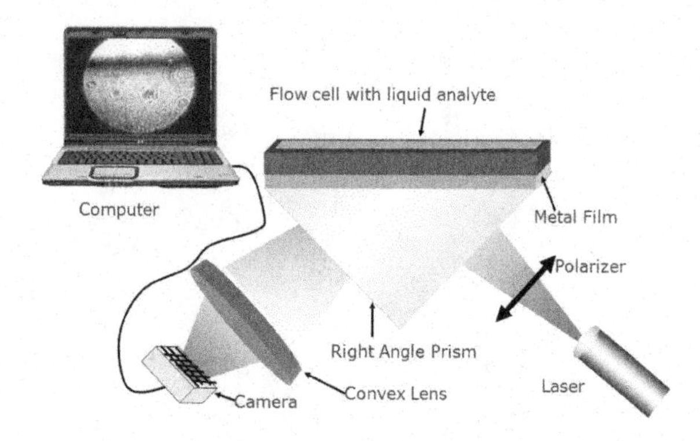

Fig. 2.20. Schematic of a diverging beam SPR setup.

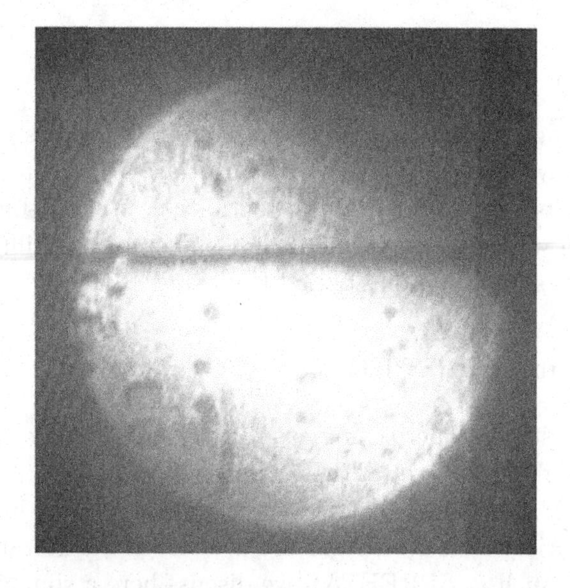

Fig. 2.21. SPR image recorded on a dieverging beam SF11/Ag-40 nm/Air setup.

resoance line. Such sensors provide cost-effective and miniaturized probes with high sensitivity.

2.9 Summary

In this chapter, we have first derived the expression for the propagation constant of the SPs and discussed the field profile, propagation length and the penetration depth. Then we have discussed the issues related to the excitation of SPs and presented three geometries of excitation. After that, the plasmon modes of a thin metal film have been discussed, where we have explained the LRSPs and SRSPs. After that, we have presented four ways to read the SPs (interrogation methods) and utilized them for sensing. In the last, we have presented the schematic of SPRI and its benefits.

References

1. A.K. Sharma, R. Jha and B.D. Gupta, Fiber-optic sensors based on surface plasmon resonance: a comprehensive review, *Sensors Journal, IEEE* **7** (2007) 1118–1129.
2. C.J. Powell and J.B. Swan, Effect of oxidation on the characteristic loss spectra of aluminum and magnesium, *Physical Review* **118** (1960) 640–643.
3. R.H. Ritchie, Plasma losses by fast electrons in thin films, *Physical Review* **106** (1957) 874–881.
4. A.K. Ghatak and K. Thyagarajan, *An introduction to fiber optics*, Cambridge University Press, 1998.
5. W.L. Barnes, A. Dereux and T.W. Ebbesen, Surface plasmon subwavelength optics. *Nature* **424** (2003) 824–830.
6. H. Raether, *Surface Plasmons on Smooth and Rough Surfaces and on Gratings*, Springer, Berlin Heidelberg, 1988.
7. S.K. Srivastava and B.D. Gupta, Fiber optic plasmonic sensors: past, present and future. *The Open Optics Journal* Bentham (2013).
8. E. Kretschmann, Die bestimmung optischer konstanten von metallen durch anregung von oberflächenplasmaschwingungen. *Zeitschrift für Physik* **241** (1971) 313–324.
9. A. Otto, Excitation of nonradiative surface plasma waves in silver by the method of frustrated total reflection. *Zeitschrift für Physik* **216** (1968) 398–410.
10. P. Berini, Long-range surface plasmon polaritons. *Advances in Optics and Photonics* **1** (2009) 484–588.
11. D. Sarid, Long-range surface-plasma waves on very thin metal films. *Physical Review Letters* **47** (1981) 1927–1930.
12. A. Shalabney and I. Abdulhalim, Figure-of-merit enhancement of surface plasmon resonance sensors in the spectral interrogation. *Optics Letters* **37** (2012) 1175–1177.
13. H. Shibata, H. Suzuki, M. Sugimoto, Y. Matsui and J. Kondoh, Development of dual-led fiber optic surface plasmon sensor for liquid refractive index detection, *Proceedings of SPIE* 6377 (2006) 6377G.

Chapter 3

Characteristics and Components of Fiber Optic Sensor

After understanding the physics of plasmons and their properties which can be used in different configurations and modes of operations for sensing applications, we are now capable of further understanding the fiber optic sensors based on them. A fiber optic plasmonic sensor comprises various components which have their own characteristics, and affect the performance of the sensor in several ways. For example, changing the doping concentration in the fiber core, or the metallic film, or the receptor layer in the fiber optic sensor changes the response of the sensor drastically, which reflects as a change in sensitivity, detection accuracy, spectral range of operation, dynamic range of sensor, limit of detection, shelf life, repeatability, reusability, and other significant characteristic properties of the sensor. To develop the understanding of sensors, it is a must to learn about various components of a sensor in general and their uses in various methods of sensing. In this chapter, we shall first present the components of a sensor and various transduction methods in general. After that we shall present a brief overview of optical fibers. Further, we shall discuss these components, their characteristics, and their significance in the fiber optic sensors. Finally, we shall discuss about the performance parameters of the fiber optic sensors.

3.1 Components of a Sensor and Their Functions

In general, a sensor has four main parts: analyte, receptor, transducer, and detector. An analyte is either a parameter or a molecule which is to be detected or sensed by the sensor. A receptor is the sensory surface immobilized with sensing elements which receives the surrounding stimuli, produces an informative signal and passes this signal to the transducer. The transducer is the third part of the sensor which converts one form of energy into another form of energy, i.e., it converts a chemical signal to an optical/electrical/acoustic signal. Fourth and last part of the senor is a detector which detects signals such as electrical, chemical, optical, acoustic, images, etc. The detector is connected to the device used for receiving final output data. These four components of a sensor can be studied under many subsections. We shall discuss these in detail in the coming sections. An illustration of a biosensor is shown in Fig. 3.1.

3.1.1 *Analyte/sample*

Chemical, biological, or environmental elements that need to be sensed are called analytes/samples. These could be natural hazards, blood/urine analytes, environmental pollutants, industrial/chemical/biochemical/biological waste, warfare agents, ambient conditions, gases, liquids, solids, endocrine disruptors, pathogenic

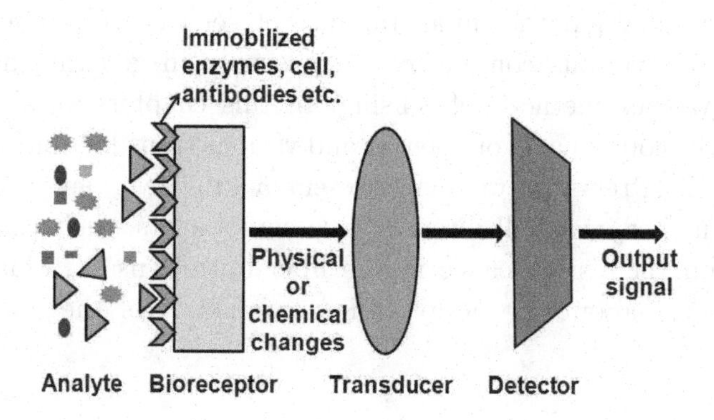

Fig. 3.1. Schematic of biosensor design.

bacteria, or other food related issues, etc. For example, some chemical vapors and gases which commonly need to be sensed are: hydrogen, oxygen, ammonia, carbon dioxide (CO_2), hydrogen sulfide (H_2S), chlorine (Cl_2), ethanol vapor, methane, etc. Hydrogen gas is used in fuel cell, engines, and also in glass and steel manufacturing plants. The most important use of this is in refining of petroleum products. Therefore, it is important to check how much gas or what concentration should be used without causing harm to the living body and environment. Particularly, the level of purity of the hydrogen gas should be checked for a variety of applications. Ammonia is another colorless gas which has a pungent smell. It is used in large quantities for refrigeration, as a cleaning material, in fertilizers, and pharmaceuticals. The flavor of ammonia is bitter and poisonous, so the optimal level of this gas should be very confined and should lie within certain limits for every use. CO_2 is a colorless, odorless gas, soluble in water. CO_2 is the natural product in the atmosphere from all the living beings using aerobic respiration and by the burning of coal, natural gases, oil, and industrial processes. CO_2 is also consumed in many foods, drinks, and packaging, i.e., carbonated beverages, freezing, and cooling. It is used in catering, metal fabrication applications, in medicine, etc. Since CO_2 is increasingly used in our day today life, it is again necessary to recognize it and its quantity for every purpose. Cl_2 is found naturally in the atmosphere and is a highly reactive gas. Cl_2 has several applications in organic and inorganic synthesis, metal manufacturing, in the cleaning of drinking water, swimming pools, sewages, pulp and paper production, etc. It is also used in medicines and pharmaceuticals. Moreover, it is important to use Cl_2 with precautions and amounts beyond certain levels of consumption are a point of concern. Hence, recognition and detection of such a gas makes sense. Ethanol is another compound of concern because it is mostly used in biofuels and medicines. If it is used for biofuel, it should be pure up to 93%,[1] while for the pharmaceuticals and organic chemistry, it should be as pure as possible up to 100%. Therefore, all the impurities in ethanol and its concentration should be known. H_2S gas is very hazardous and gets easily absorbed by the lungs, which can create irritation in skin, eyes, and nose. It can cause asthma, smell disability, headache, and insomnia. Therefore, the sensors to detect the small concentrations of H_2S must be employed in mines, sewages, and also in air monitoring equipments.[2] A number of other gases and vapors exist in the atmosphere, out of those, some are good and some are bad, while for some cases particular concentrations are useful but low or high concentrations could be hazardous.

If we look at the ambient conditions (pH, temperature, humidity, pressure), they also play crucial roles in harvesting, manufacturing of medicines, beauty products, polymers, paints, and other relevant activities/commodities in daily life. For example, pH should be maintained in the swimming pool. It is also very important to measure and balance the pH of the blood at regular intervals. Humidity levels should be maintained for the patients in anesthesia and those who are undergoing critical care medication.[3,4] Several chemical and biological compounds, i.e., metal ions, hydrogen peroxide, glucose, urea, cholesterol, vitamins, ethanol, tetracycline, phenol compounds, pesticides, naringin, melamine, cocaine, dopamine, potassium, etc. are needed to be sensed, which are critical for food and medical diagnostic industries.

It is very important to maintain an optimum level of metal ions in the drinking water. The levels of dissolved compounds in drinking water should be optimum. Some metals/compounds such as lead, arsenic, iron, lithium, beryllium, polonium, bismuth, cadmium, mercury, etc. can be potentially dangerous, if found in higher amounts in drinking water. Hence, the measurement of pollutants in water is quite critical.[5] Moreover, the level of metal ions should be maintained through several purification processes. Sensors for the detection of metal ions are important for both the contaminated and purified water. Further, tetracyclines are the kind of antibiotics used in harvesting for the protection of corps, vegetables, aquaculture plants, and in the farming industry for the germs protection. Moreover, tetracyclines may remain as residuals beyond the permissible limit in food stuff.[6] A certain level of tetracycline is acceptable but it will be harmful for the body beyond its permissible residual limit. For example, the level of concentration of tetracycline in milk should be 0.1 mg/kg and in honey, 0.01 mg/kg. Glucose is an important analyte of human blood which has got the most attention of researchers for diagnostic purposes. It has attracted huge interest as higher amounts of glucose in blood leads to diabetes. Similarly, lower blood glucose levels lead to several brain related diseases such as unconsciousness and neuroglycopenia.[7] The amounts of sugar in blood are maintained by insulin. Sometimes, the disorders of insulin secretion lead to imbalance in blood glucose. Hence, certain sensors for insulin have also been in practice. Similarly, urea is another main component of human blood and urine. Amounts of urea out of the physiological range may take part in the malfunctioning of kidney and other organs. It is also used for fertilizers in the harvesting, clinical chemistry, and environmental monitoring.[8] Therefore, urea residues in crops should be in healthy limits. Cholesterol is a fatty substance which constitutes the lipids in the human blood. Four different forms of cholesterol; intermediate density lipoprotein (IDL), very low density lipoprotein (VLDL), low density lipoprotein (LDL), and high density lipoprotein (HDL) are found in human blood. Similarly, triglycerides are other forms of lipids. The level of all cholesterol and triglycerides should be maintained within certain physiological range to avoid the possibilities of cardiovascular and heart diseases.[9] Apart from these compounds, hemoglobin, iron, sodium, potassium, creatinine, blood cell counts, etc. are also important analytes which need to be checked regularly and maintained in the normal physiological range in blood. Other important blood analytes are vitamins. Vitamins are a group of organic substances, present in minute amounts in natural food stuffs that are essential for normal metabolism, growth, and development of the human body. Insufficient amounts in the diet may cause deficiency diseases. Vitamins are characterized in 14 varieties: A, B_1–B_7, B_{12}, folic acid, C, D, E, and K. These vitamins are obtained through the daily food and every vitamin has its own functions and specialty. We shall explain a bit more about some of the important vitamins, their sources and the utility of their regular monitoring. Vitamin C is found in fresh vegetables, lemon, cabbage, green and red pepper, etc. It maintains health of the body parts such as skin, teeth, blood vessels, etc.[10] Deficiency of vitamin C leads to scurvy disease, bleeding gums, brushing, and loss of teeth, etc. Further, vitamin A has its own importance for several functions in human body such as eye vision, reproduction, and cellular communication. It also helps in the cell growth and differentiation and in the maintenance of kidney and lungs. Vitamin B is a group of water soluble vitamins; it contains nine sub parts. They play important roles in cell metabolism and other functions in the body. Presence of every vitamin of this group supports several body parts and their functioning. For example, vitamin B_{12} plays a key role in the formation of the blood for the brain and nervous system. It is involved in regulating metabolism and the cells in the human body and in energy production. Vitamin B_5 is an essential nutrient for many animals and it takes part in the metabolism of proteins, carbohydrate and fats.[11–13]

There are several pesticides, antibiotics, minerals, and other compounds of interest for chemical and biosensing and they are usually needed to be sensed/detected for a balanced daily life. These analytes are sensed in complex media such as blood without the interference of other constituents. The sensors for specific detection of these analytes are generally based on biomolecular recognition elements (BREs) called *receptors* specific to the analyte of interest. These receptors could be enzymes, aptamers, nucleic acids, microorganisms, tissues, cells, antibodies, etc. The recognition elements have one of the following: fluorescence properties, electro optical capabilities, mass sensitive, thermo optic sensitive, magnetic properties, and ion selective. Every sensor works on a specific principal which depends on the particular property of the analyte. Therefore, sensor and its fabrication, working principal, and whole dynamics depend on the analytes of interest. As pointed out earlier there are a number of recognition methodologies hence, we shall discuss now the working principles/properties of them thoroughly in the subsequent sections.

3.1.2 *Receptors*

A receptor can be defined as the heart of a biosensor. It is designed for the interaction of the analyte with it to produce a significant signal which is to be measured. A receptor has to be selective and specific to interact with the biomolecule/analyte of interest. The most desirable receptors are the ones which are able to detect the analyte correctly and specifically in complex media, without the interference from other similar agents. The immobilization of biomolecule over surface and creation of binding sites depends on the type of the molecule used as a recognition element and on the analyte. The receptors are of many varieties according to different bio-recognition events that have been used for monitoring. Most of the physical sensors such as fiber Bragg gratings, temperature sensors, pressure sensors, etc. generally do not require any receptors and mostly depend on the physical properties of the materials used in the fabrication of the sensor. However, most of the chemical/biochemical sensors and those important for medical diagnostics utilize receptors due to the issues related to specificity and selectivity. Generally, receptors can be classified into five major categories according to the interaction taking place: (i) enzyme based, (ii) antibody/antigen based, (iii) nucleic acids, (iv) cell based, and (v) tissue based.

3.1.2.1 *Enzymatic receptors*

Enzymes are basically the macromolecular proteins which act as biomolecular catalysts. They are very important for life because they are highly responsible for the metabolic processes, such as reproduction, growth, development of organs, digestion, etc. Enzymes are very specific catalysts that accelerate the reaction rate as well as provide the uniqueness of the metabolic reactions. Several reactions in the human body, i.e., digestion of the food, synthesis of DNA, etc. are the enzymatic reactions. They control the kinetics of a particular reaction among a number of simultaneous reactions due to other processes and enzymes. Further, enzymes give a right pathway to metabolic reactions which generate cells in the human body. A number of very famous enzymes such as *urease* for the conversion of urea into ammonia and CO_2, *glucose oxidase* (*GOD*) for the catalysis of the reaction of glucose with oxygen producing gluconic acid and hydrogen peroxide (H_2O_2) and *chlolestrol oxidase* for the detection of cholesterol[9,14,15] have been repeatedly used in sensing. There are several enzymes that work with cofactors and/or coenzymes in catalysis process such as *alcohol dehydrogenase* (*ADH*) working with nicotinamide adenine dinucleotide (NAD).[16] Enzymatic reaction is also affected by the different molecules; some of them decrease the enzyme activity called inhibitors, while some others increase the activity called activators or promoters. Enzymes are substrate specific as well, and hence cannot work with every substrate. For example, urease gets inhibited by metals and hence cannot be used directly with metal substrates unless the active site is blocked and attached to the surface from other sides by different means. Hence substrates always need to be chosen accordingly.

Enzymatic receptors have to interact with various analytes at particular environmental conditions. The chemical reaction results in a reaction product and the enzyme become separated from the analyte. The reaction product gives measurable signals which make a difference in the detection process. Certain chemical reactions occur during enzymatic reaction process. A specific enzyme works with specific compounds and quantity and quality of enzyme can accelerate and decelerate the chemical reaction. Enzymatic reaction also depends on the pH of the chemical compound, hydrophobicity, concentration of the chemical compound, and sample solution. Obviously, enzymatic reaction also depends upon the ambient conditions, i.e., temperature, humidity, pressure, etc. The enzymatic reactions are reversible hence the enzyme surface can be regenerated and used repeatedly for certain times. This feature is advantageous in terms of the reusability of the sensor.

Most of the enzymes are proteins except ribonucleic acid molecules and those which work with their amino acid residues.[17] Some enzymes need some other chemical groups to activate the reaction. The chemical groups are called *cofactor* of the enzymes. These kinds of chemical groups could be inorganic ions Fe^{2+}, Mn^{2+}, Mg^{2+}, and Zn^{2+}, complex organic, and/or metallo-organic molecules.[18] Recognition capabilities of the enzyme-based receptors may be destroyed by denaturing and dissociation of the enzyme and breaking them down into amino acids. The enzymes at the enzymatic receptor surface lose their activity after multiple usages, so they need to be replaced periodically. Enzymatic reactions may also be affected by other red-ox active reactions. Hence enzymes work productively, with sophisticated chemical reactions.

There are a lot of enzymes in use for biosensing. GOD is the first enzyme which was used for the biosensing for the detection of glucose by Clarke.[19] *Glucose oxidase* is able to give information about glucose concentration by one particular oxidation reaction. There are few more examples of enzymes used in biosensing, i.e., *urease* for the detection of the urea, *oxidoreductase* for the detection of lactate,[20] *lyase* for citric acid analysis,[21] etc. A list of enzymes that are used for particular analytes is given in Table 3.1. The preparation of the enzyme generated receptors is simple, inexpensive, and quick. These receptors are reusable and the performance capability of these receptors also depends on the enzyme loading, thickness of the enzyme layer, enzyme immobilization process, its coenzymes and cofactors. The binding capabilities given by the enzymatic reaction are less than other binding techniques.

Table 3.1. List of enzymes and corresponding analytes.

S. No.	Enzyme	Analyte
1	Catalase	Sucrose, glucose
2	Lactate dehydrogenase	Lactate
3	Alchohol dehydrogenase	Ethanol
4	Glycerol kinase	Glycerol
5	Choline dehydrogenase	Phospholipids
6	Lactate oxidase	Lactate
7	Mutarotase	Glucose
8	Sulfite oxidase	Sulfite
9	Urate oxidase	Uric acid
10	Penicillinase	Penicillin
11	Lutamatedehydrigenase	L-glutamic acid
12	L-Tyrosine decarboxlase	L-tyrosine
13	Glutaminase	L-glutamine
14	Glucosidase	Amygdalin
15	Nitrate reductase	Nitrite and nitrate

3.1.2.2 *Antibody-based receptors*

Antibodies are proteins having different shapes and sizes. They are an essential part of the human body and are produced by the plasma cells. Antibodies are used by the immune system of the human body because they are capable of recognizing the foreign objects and binding them. These foreign elements are called antigens, for example, viruses and bacteria. Antibodies act as a marker and indicate to other parts of the immune system to attack the foreign elements, once captured by them; these dangerous molecules are removed from the surrounding and thrown out of the body. All the antibodies have almost the similar structure except a very small part at the sites capable of binding to the antigens. That small binding site at the top is called epitope which binds with the paratope of the antigen. Billions of antibodies flow in human blood and recognize, and fight with different diseases.

Antibodies are grown or developed by the immune system itself, when an unknown antigen is detected. In general, the antibodies are grown in some animal species by inserting the antigen into the animal's body (immunization) for some time for the development and then taken out and purified by several steps. The size of the animal (actually the blood vessel) is chosen according to the desired amount of the antibody. These antibodies can be used as bio-recognition elements, as they are capable of interacting with certain antigens. Such a sensing is named as *immunoassay*. Antibodies are immobilized over any support and/or substrate to make it a receptor to receive signals from a specific antigen (analyte) and interact with them to produce output signals. There are primarily two kinds of antibodies which are chosen according to the antigen or analyte type. These different antibodies are described next.

3.1.2.2.1 Polyclonal antibody

Polyclonal antibodies are a population of antibodies that can recognize multiple epitopes on one antigen. When an animal is immunized with an antigen, it elicits a primary response of the immune system. That is followed by secondary, tertiary, and further immunizations, which lead to higher titers of the antigen. When the immune system responses to the antigen, it creates the involvement of multiple B cells and these B cells are the actives for specific epitopes of that particular antigen. At the end of this process the obtained serum contains a heterogeneous complex mixture of polyclonal antibodies of different affinities for different epitopes of a specific antigen. These antibodies are mainly produced in rabbits, goats, or sheep. These are specially chosen for pathogen detection. Most of these antibodies are used in sandwich assays, as they can bind at different epitopes of the antigen. They also provide the possibility of tagging with labels, which find application in enzyme linked immunosorbant assay ELISA and fluorescence spectroscopy. Mostly polyclonal antibodies are used for serological purposes and immunohaematology.

3.1.2.2.2 Monoclonal antibody

Monoclonal antibodies are made from a single immune cell and hence the clone of a single parent cell unlike the polyclonal antibody. Monoclonal antibody has the ability to bind with only one epitope of the antigen. Monoclonal antibodies are very specific and they decrease the effect of unwanted signals and noise. However, the production of these antibodies is expensive.

For the antibody-based receptors the first thing is to select the antibody according to antigen/analyte to be sensed or detected. Afterward the suitable immobilization technique is decided. We have described different immobilization techniques in Chapter 5. The antibody-based receptors are the traditional immunosensors for the detection of samples/analytes in a sub nano/picomolar concentration range.[22] When these antibodies come in contact with the antigens or analytes, they bind at the epitope of the antigen. These interactions are based on the immunochemical reactions. The interaction of the antibody with antigen leads to the formation of antigen–antibody complexes that make changes in measuring signal at the output. Antibody–antigen binding is strong enough so that other side-interactions get minimized. For achieving good sensitivity of antibody-based sensors, some enzymes are required for the making of strong interactions. In these sensors, receptor is immobilized with antibody along with enzymes. These types of receptors operate in equilibrium and they could afford a fixed amount of supporting enzyme. This type of reaction is called ELISA.[23] In general, five kinds of antibodies are found in a body: IgG, IgA, IgM, IgD, and IgE, where Ig (immunoglobulin) is the abbreviation for antibody. These antibodies have different functions and are specifically used against various antigens. The antibody–antigen interaction depends on the physical and chemical states of the antigen and antibody and also on the environmental conditions. This reaction also depends on the concentration ratio of the antibody and antigen and it can lead to dissociation of the antibody–antigen complex. The antibody–antigen reaction is a reversible reaction but when these are coupled with enzymes, (i.e., in ELISA) then it becomes an irreversible reaction. For example, antigen can be detached from the antibody by changing the pH conditions. Antibody–antigen interaction based sensors are mostly used for the diagnosis of the viral, fungal, bacterial, and parasitic infections; to monitor the immunity of the human body and cause of the infection.[24,25] However, they are highly used for laboratory and research purpose diagnosis.

3.1.2.3 *Nucleic acid based receptors*

Nucleic acids are large biomolecule elements, essential for every living creature. Nucleic acids contain dioxyribonucleic acids (DNAs) and ribonucleic acids (RNAs) known as nucleotides. They also include peptide nucleic acids and aptamers. Aptamers are the oligonucleic acids, which have different shapes and sizes and can be selected by a series of random oligonucleotides to bind with several molecules of biomedical interest. Nucleic acids are relatively new, important, and exciting bio-recognition elements for the monitoring of a variety of parameters in different fields such as the food industry, clinical diagnosis, forensic sciences, environmental protection, etc. DNA and RNA are highly used for clinical diagnosis, cure of several diseases, and detection of pathogenic infections.[26,27] The basic principle of a nucleic acid receptor based sensing lies on sequence complementarities of base pairing (for DNA, i.e., A, T, G, C), although it is not always true for aptamer-based devices. However aptamers can be used as BREs in sensors utilizing optical bioassays, as micromolecular drugs in basic research and for clinical purposes.[28] They can further be used to regulate cellular processes and to guide drugs for the detection of several biological elements, i.e., protein, cells, bacteria, viruses, organic dyes, etc. Aptamers have good binding ability and a less denaturation and degradation possibility. Moreover, they can be used as chiral selectors that can differentiate between chiral molecules.[29]

Nucleic acid based receptors are formed by the immobilization of the nucleic acids over solid support/base with appropriate techniques. Immobilization always depends upon the support and bio-recognition element. DNA-based receptors can be prepared by appropriate chemical methods but the RNA-based receptors can be made by reverse transcription of previously isolated or particular messenger RNA or its nucleotide sequence based on the amino acid sequence. Out of these two, DNA is easy to synthesize and forms a layer over the support matrix. Further, it is highly stable and reusable. Nucleic acid receptors work for the detection of several biological analytes with electrical/chemical/optical transducers. The interaction of the nucleic acids with target/analyte generates optical detectable signal (in the case of optical transducer) that helps to get the information about the target. Nucleic acid based receptors are very advantageous over enzyme and antibody based sensors as per the earlier mentioned qualities. For example, DNA-based receptor is capable of amplifying the desired target signal from the host pathogens. Nucleic acid based receptors can be employed for the multi analyte detection that reduces the time of analysis, cost, and undoubtedly labor as well.

3.1.2.4 *Cell-based receptors*

Cells, called the building blocks of life, are the smallest units of life and have their own structures and functions. The receptors utilizing cells are generally based on the binding of the target molecule with entire cell/microorganism or a particular cellular component that is immobilized over the receptor surface. The receptor responds to the target compounds and generates a measurable signal. The principal of the receptor is based on the consumption of specific chemicals into cells for acquisition. Because of the consumption of specific chemicals, they are compatible with particular analytes only. Hence they have very good specificity, i.e., bacteria and fungi are used as the indicators of toxicity. In a few examples of the cell-based receptor, cell metabolism and cell respiration are used for the detection of toxicity of the heavy metals in drinking water and other edibles.[30] Cell-based receptors are long lived because the reaction between the cell and analyte is a closed system and are immune to the surroundings. The mammalians cells as well as plant cells have been used for biosensing applications.[31] The cell-based biosensors have been used for the detection of short chain fatty acids in milk. *Arthrobacter nicotianae* microorganisms have been used and immobilized over the support with calcium alginate for sensing. The oxygen consumption quantity of the *A. nicotianae* is measured as an output signal.

Cell-based receptors have a very low limit of detection due to the signal amplification, which is very advantageous over other receptors. These receptors also work on the catalytic or pseudo catalytic properties of the cells. Cell-based sensors can be used for the analysis of the effect of pharmaceutical compounds.[32] Proteins found in cells can also be used for the monitoring of various targets.[33] They can be used in the biomedical and environmental interest for the monitoring of pathogens, toxins, etc. and in the pharmaceuticals industry and in drug detection.[34–36] Moreover, they can also perform real time monitoring of bioassays dynamically and rapidly and they are inexpensive too. Apart from these good factors, they have certain limitations also such as, the issues related to the stability of cell, which depends on many factors (sterilization, life time, etc.). Hence, these sensors are not very stable. Moreover, cell-based receptors are sometimes less selective because of the multi receptor behavior of the intact cell over the support.[37] In some exceptional cases, for example, the sensors performing bacterial detection based on specific bacteriophages provide better stability and specificity over antibody and other cell-based receptors.[38]

3.1.2.5 *Tissue-based receptors*

Tissues are the organization of the cells which perform similar functions. They look like a lattice of similar cells from the same origin. A group of multiple tissues forms organs, which further form living creatures. Tissues form the body of every living creature; be it a human, animal or a plant. In general, there are four types of tissues in living body. First is the connecting tissue. It stores nutrients, cushions, protects organs and joints, and connects muscles to bones, and gives strength to the skin. Second is the epithelium tissue that protects us from the outside world, absorbs water and nutrients, works as a filter in the kidney, and eliminates waste products. Third is the nervous tissue that conducts impulses to and from the body organs via neurons. Fourth is the muscle tissue. These are responsible for the body movement, blood movement, and waste food and mechanical digestion. Moreover, plant tissues are of several kinds: Meristemetic tissue takes part in dividing the cells and increasing the length of the plants; protective tissues cover the surfaces of the leaves, roots, and stems. The xylem tissue works as a feeder, conducts water, and distributes minerals from the root to other parts of the plants; while the phloem tissue is the companion tissue and brings the food from the leaves to the storage parts of the plant and examines many parts. These tissues are in use as sensing elements in the field of biosensors and biomedical diagnostics.

Tissues are used as bio-receptors because they consist of a large number of enzymes. They have essential cofactors to interact and react with the analytes and function to give a reliable signal. For example, the banana tissue has been used for the sensing of dopamine.[39] Banana tissues contain a high quantity of polyphenol oxidase (PPO) that can be mixed with a carbon matrix. The combined matrix has good detection capabilities for dopamine. They are easy to immobilize over the receptor, highly active, low in cost, and have good selectivity. Although they have lots of advantageous features, they do not have good stability due to the presence of unwanted enzymes with them, which badly affect the biological reaction and give equivocal output signals. Tissue-based receptors are also used in defense because they can easily and accurately detect the biological and chemical weapons.[40]

3.1.3 *Transducers*

The third and most important part of a sensor is a transducer. The very basic definition of a transducer is that the device can convert one form of energy into another form of energy. Hence, it can transform the input/chemical energy change occurring at the receptor surface into a measurable output signal. In other words, a transducer provides a relationship between output and the given input. The activity of the analyte over the receptor (interaction between bio-recognition element and analyte) can be monitored by the production of hydrogen peroxide, change in current, consumption of oxygen, change in resonance signal, fluorescence, mass, temperature, pH, absorption, acoustic signal, intensity, phase, image, etc. Hence in context to sensors, the transducers can be classified according to the working principle of the device and can broadly be divided into following four categories:

- Electrochemical
- Piezoelectric
- Thermometric
- Optical

A chart presenting the classification of transducers in context to sensing applications is shown in Fig. 3.2. We shall discuss these transducers in detail in the subsequent sections.

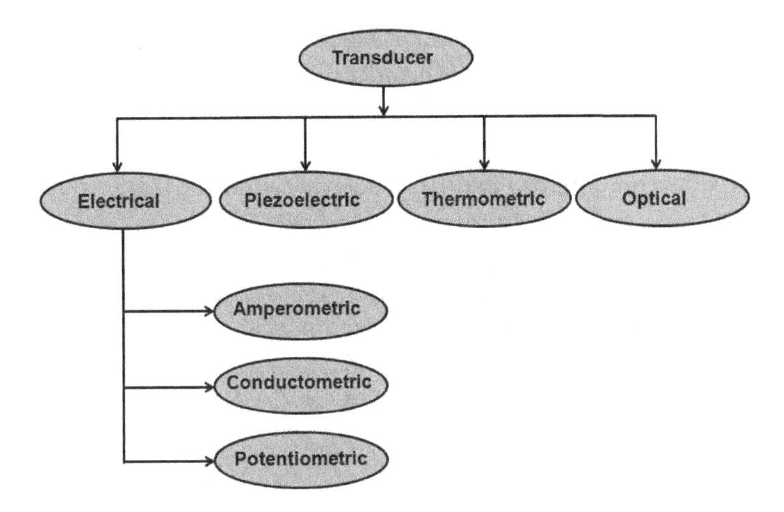

Fig. 3.2. Classification of transducers in context to sensors.

3.1.3.1 *Electrochemical*

An electrochemical transducer is an integrated device which provides a measurable quantitative account of the modulations in the input electrical signal occurring due to the interaction between receptor and the analyte. Sometimes, the interaction can directly be transformed into electrical signals without any input signal. Electrochemical transducers utilizing the chemical interactions are called electrochemical transducers, which convert the chemical energy into electrical energy and vice versa. The reaction occurs because of the electron transfer or by a potential scan. Electrochemical transducers are basically used for the sensing of DNA, glucose, etc. Electrochemical sensors contain three electrodes: the first being the working electrode; second, the reference electrode and the third, auxiliary electrode. The working electrode is generally made of chemically stable materials such as gold, platinum, graphite, etc. Most commonly, the reference electrode is made of silver (Ag) coated with silver chloride (AgCl) and the auxiliary electrode is made of platinum. The interactive reaction between the target analyte and the receptor occurs at the working electrode. Some electrochemical transducers have only two electrodes: the working electrode and the reference electrode.[41] Both types of transducers are used for sensing. The three electrode transducer system is more stable and long lived as compared to two electrode transducer system. However, two electrode transducer system is used in disposable sensors. Further, the two electrode transducers are inexpensive as the long life reference electrode is not required. Electrochemical sensors can further be classified in following three subsections according to the measurable electrical variable: (i) amperometric (ii) conductometric, and (iii) potentiometric. These are discussed briefly as follows.

3.1.3.1.1 Amperometric

When a constant potential is applied across the working electrode and reference electrode then electrochemical oxidation or reduction generates a measurable current which is also known as voltametric. It is another form of the amperometric detection and mostly used as a laboratory technique. The generated current is proportional to the number/amount of analyte (chemical/biological species) present at the receptor surface. Amperometric biosensors are very sensitive and fast because they can participate in red-ox reactions and can transfer electrons very rapidly[42] which makes changes in chemical reaction and hence produces the changes in the current. These biosensors are highly in use because they work as biochemical reaction mediators.

3.1.3.1.2 Conductometric

An interaction between the analyte and the receptor may result in the change in the conductivity of the medium between electrodes. This change in conductivity can be measured through conductometric transducers. The conductometric transducers are capable of measuring changes in electrical conductivities of the analyte, composition of the sample, and/or any medium like nanowires, etc. These sensors include enzymatic receptors very often. The interaction between the analyte sample and enzymatic receptor causes the change in ionic strength, which reflects an increase/decrease in the conductivity.

Conductometric sensors are mostly used in environmental monitoring and clinical analysis. Conductometric biosensors have been developed for the detection of *Salmonella* and *E. coli* O157:H7, which are very important for biosecurity.[43] These transducers have also been reported for the detection of methamphetamine drug in human urine.

3.1.3.1.3 Potentiometric

In this type of transducers, analyte–receptor reaction results in a measurable output signal as change in potential or charge accumulation of an electrochemical cell at negligible current flow. This electrochemical transducer is made of one glass pH selective electrode and another ion selective electrode such as Na^+ Cl^-.[44] The transducer works on the measurement of the potential difference between two ion-selective membranes. The transducer can be coated with any biological element to use it for biosensing. A penicillin sensor has been developed by the coating of a pH electrode by *penicillinase*, which works as a catalyst for the reaction between penicillin and *penicillinase* coated electrode and generates H^+ ions. As the H^+ ions create changes in the pH of the electrode, this change is translated into a measurable change in potential and hence, penicillin.

3.1.3.2 *Piezoelectric*

The piezoelectric transducer has a piezoelectric crystal that plays main role for the detection of analytes. Piezoelectric transducers are based on the coupling of biological/chemical sample/analyte to a piezoelectric crystal. Several piezoelectric crystals such as quartz, lithium niobate, tourmaline, aluminium nitrate are in use. Frequency of the oscillation of the crystal depends on the frequency applied to the crystal and also at the crystal mass. Piezoelectric transducers such as quartz crystal microbalance (QCM) are highly mass sensitive. Application of the analyte over the receptor causes the binding of biomolecules over the crystal which makes a change in the mass of the crystal which leads to a change in the vibration frequency and hence a change in the output measurable signal. At last, the total change in the mass of the crystal is detected by translation of change in the frequency in it.[45] These transducers are convenient and can be used for real time monitoring and label free detection. In addition, these transducers are highly in use for the molecular detection, environmental, food, and clinical diagnosis. For example, piezoelectric biosensors have been employed for the detection of hepatitis B, hepatitis C, and food borne pathogens.[46,47] Moreover, it is found that piezoelectric transducers are capable of detection up to concentrations of 8.6 pg/L of hepatitis B and 25 ng/L of cholera toxin[48] which is advantageous for biosecurity.

3.1.3.3 *Thermometric*

Thermometric transducers are based on the thermostat or thermopile. The basic idea behind a thermometric transducer is the measurement of the consumption and production of heat during the interaction of an analyte with the bio-recognition element. In this case the transducer monitors the energy produced by the heat transformation during the receptor–analyte interaction. The change in heat/temperature can be correlated with the concentration of analyte, type of analyte, and the involved bio-recognition molecule. The chemically generated signals, in general, do not have complete information about the analyte because of some heat losses in irradiation, conduction, or convection. These transducers are not good enough because they give incomplete information about the analyte. These transducers do not have good sensitivity but they are stable and have good possibilities of miniaturization. First calorimetric transducer based immunosensor was developed for a cell by using enzymes.[49] Calorimetric transducers have been employed in food industry, cosmetics, and environment studies.[50–52] Several thermometric sensors have been developed for the detection of various biological and chemical compounds, such as glucose, lactate, penicillin G, oxalic acid, etc.[53]

3.1.3.4 *Optical transducers*

Optical transducers measure the characteristics of light as output signals. These are, generally, based on various optical properties/principles such as fluorescence, luminescence, optical diffraction, evanescent wave, and surface plasmon resonance (SPR), etc. The interaction of the receptor with the specific analyte results in changes in the mass, concentration, properties, and number of molecules on the receptor or in the analyte. This causes a change in the characteristics of light (intensity, wavelength, phase, wave-vector, etc.) hence the output signal. Fluorescence-based transducers detect the change in electromagnetic radiation after chemical reaction between the analyte and recognition element. There are several fiber optic chemical and biosensors based on evanescent wave, luminescence, and SPR. At present, optical transducers are found the most reliable because they have many advantageous features over other transducers; i.e., high sensitivity, safety, simplicity, miniaturization, immunity to electrical shocks and portability. Further, fiber optical transducers can be used for remote sensing and online monitoring.[54] Apart from these qualities, they have few drawbacks as well. To quote a few, they do not have good stability and selectivity. Sometimes recognition elements/compounds are very sensitive to ambient light that can generate false signals.[54] Optical transducers have been employed for the detection of nucleic acids, pathogens, viruses, etc.[55]

3.1.4 *Detector*

Detectors are the fourth and the last unit of a sensor. They are selected according to the transducers. A detector could be any current measuring device, thermometer, optical power meter, fluorometer, pressure measuring device, camera, spectrometer, magnetometer, etc. In principle, a detector always measures the output signal from the transducer. In optical fiber sensors based on plasmonics, the main topic of the present book, the detector is a spectrometer which measures the output optical spectrum with respect to different analyte concentrations. Here we shall give brief introduction about a spectrometer.

A spectrometer is a device which measures the intensity of the light for a range of wavelengths in a certain spectral range. A spectroscopic instrument generally consists of an entrance slit, collimator,

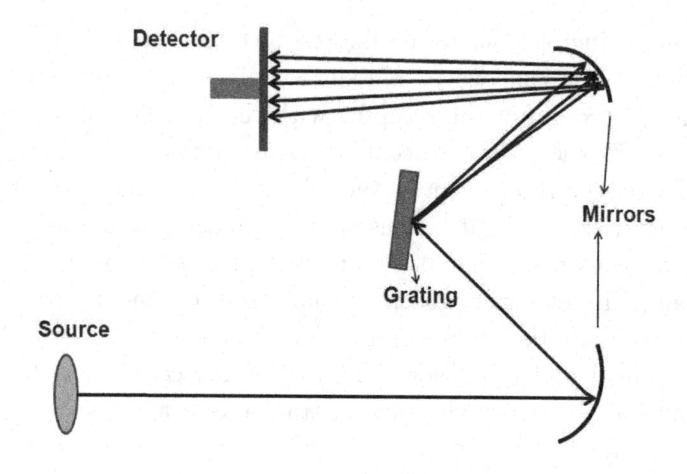

Fig. 3.3. Layout of optical bench.

dispersive element such as a grating or prism, focusing optics and a/an detector/array of detectors. The detector incorporated in spectrometer is most generally a CCD linear array of thousands of pixels, which lead to fast processing of the spectroscopic data. Spectroscopic measurements are being used in many different applications, such as colour measurements, concentration determination of chemical components, or electromagnetic radiation analysis. In the context of fiber optic sensors, we restrict ourselves to the discussion of a fiber optic spectrometer. A schematic picture of the optical bench of a spectrometer is illustrated in Fig. 3.3. The spectrometer has a fiber optic entrance connector, collimating and focusing mirrors, and a diffraction grating with this input light through optical fiber enters into the spectrometer. The spectrometer consisting of a grating has few hundred lines/mm and a blaze angle which enables applications in the desired wavelength range. In the optical bench light enters through a standard fiber connector and is collimated by a spherical mirror. A plane grating diffracts the collimated light which is focused by a second spherical mirror on the one dimensional linear detector array. The choice of these components such as the diffraction grating, entrance slit, order sorting filter, and detector coating, has a strong influence on the system specifications like sensitivity, resolution, bandwidth, and stray light. Since the data capture process is inherently very fast, the speed of measurement allows in-line analysis. The modern times spectrometers are interfaced with a computer and from where output data can be exported.

3.2 Optical Fiber

Optical fibers play a significant part in the fields of communication, sensing, and in the designing of many other devices. Optical fiber is an optical wave guide that carries information with a light beam from one end to the other end of it. Light is guided into the optical fiber by the principle of total internal reflection (TIR) (we shall discuss TIR later in this chapter). The structure of an optical fiber, in general, is cylindrical, which comprises of two concentric cylinders of different refractive indices (RI). The structure is further covered with a plastic jacket for the protection of the fiber from environmental effects. An illustration of the structure of an optical fiber is shown in Fig. 3.4. The inner cylinder is called the core of the optical fiber, while the outer one is called the cladding. The refractive index of the material of the core (n_1) is slightly greater than that of the cladding material;

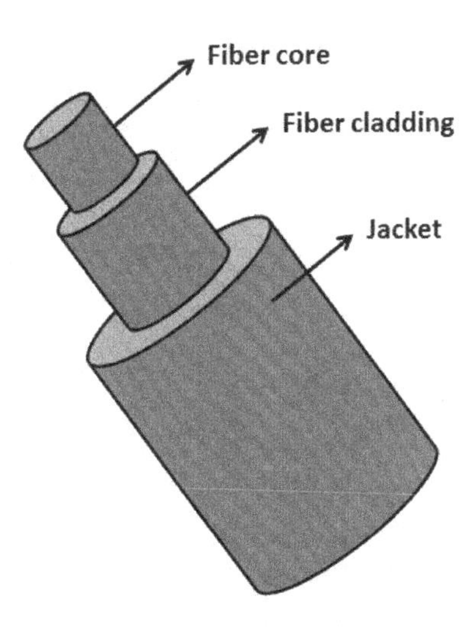

Fig. 3.4. Structure of optical fiber.

i.e., $n_1 > n_2$. Refractive index profile is given as[56,57]:

$$n = \left.\begin{matrix} n_1 & 0 < r < a \\ n_2 & r > a \end{matrix}\right\}, \tag{3.1}$$

Where r is the cylindrical radial vector (transverse direction) and a is the radius of the core which can extend up to a certain value. There are various designs of the fibers with different core, cladding dimensions, and structure. We define a parameter Δ, called fractional refractive index change, given as:

$$\Delta \approx \frac{n_1 - n_2}{n_1}, \tag{3.2}$$

The parameter Δ defines the difference between the core and cladding RI. Cladding is important to support the fiber core because at the time of the transmission, fiber needs supports any way and the supporting device or structure can damage it. To understand the guidance of light inside the fiber, TIR phenomenon is important to understand.

3.2.1 *TIR*

When a beam of light is incident at the interface of two media having RI n_1 and $n_2 (n_1 > n_2)$, at an angle θ_i then incident light gets divided into two parts; one is reflected back in the same medium at an angle called *angle of reflectance* and having the same magnitude as the *angle of incidence* (see Fig. 2.4(a)). The other part is refracted at an angle θ_r, *given by Snell's law* ($n_1 \sin\theta_i = n_2 \sin\theta_r$), in another medium of refractive index n_2. If the light is incident at an angle greater than the *critical angle* (see Sec. 2.3.1), all the light gets reflected back into the same medium where it came from. This phenomenon is known as TIR. It occurs within a range of angles, lying between the critical angle (θ_c) and 90°. Since the refractive index of the fiber core is slightly greater than that of the cladding, light incident on the optical fiber gets guided into it through TIR for a range of angles ranging from the critical angle till 90°. The rays apart from those, leak through the core-cladding interface and get dissipated within few microns of length of the optical fiber.

3.2.2 *Light ray propagation in an optical fiber*

To understand the light guiding phenomenon into an optical fiber, let us consider an optical fiber with core of RI n_1 and cladding with n_2 as the RI ($n_1 > n_2$). An illustration of the lateral view of the fiber is shown in Fig. 3.5. A ray enters the optical fiber at an angle θ_I and gets guided along the fiber with the incident angle $\theta_b(>\theta_c)$ at core-cladding interface through the repeated TIR inside the optical fiber. We have taken here a tiny piece of optical fiber to show TIR inside it, but the same phenomenon occurs for thousands of meters of long fiber because of its cylindrical symmetry. However, several losses in guided light occur due to bends (if any) of optical fiber, material dispersion, pulse dispersion, scattering, etc. A rigorous discussion of such loses is out of the scope of this book and can be found elsewhere.[58]

In Fig. 3.5 we illustrate the light guidance into a step index fiber that has a sharp change in RI at the core-cladding interface. Moreover, light ray propagation through TIR occurs in all fibers, for example, in graded index fibers as well, where RI of the core gradually decreases with increasing distance from the fiber axis.

3.2.3 *Numerical aperture*

Numerical aperture is a numeric number for an optical lens or any other device that specifies the range of angles over which light can be accepted or emitted by the device. As we have seen in Fig. 3.5, light is incident at an angle θ_I from air (RI $= n_0$) to the fiber core

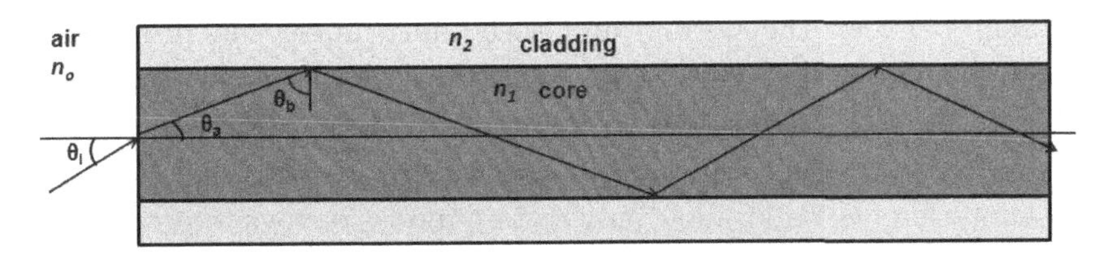

Fig. 3.5. Propagation of light ray in optical fiber.

and gets refracted at an angle θ_a. Following the Snell's law:

$$n_0 \sin\theta_I = n_1 \sin\theta_a,$$
$$\sin\theta_I = n_1 \sin\theta_a (n_0 = 1 \text{ for air}),$$
$$\sin\theta_I = n_1 \cos\theta_b, \tag{3.3}$$

for the case $\theta_b = \theta_c$:

$$\sin\theta_I = n_1 \cos\theta_c, \tag{3.4}$$

for this case θ_I is the maximum angle through which fiber can accept light along its axis. The angle $2\theta_I$ is called the *acceptance angle* through which light can enter the fiber. Term sin θ_I in Eq. (3.4) is called numerical aperture (NA) of the fiber and it can also be written as:

$$NA = (n_1^2 - n_2^2)^{1/2},$$

From Fig 3.5 $\sin\theta_I$ can be written as:

$$\sin\theta_I = n_1(1 - \sin^2\theta_b)^{1/2},$$
$$\sin\theta_I = n_1 \left[1 - \left(\frac{n_2}{n_1}\right)^2\right]^{1/2},$$
$$NA = \sin\theta_I = (n_1^2 - n_2^2)^{1/2}, \tag{3.5}$$

Larger the NA, higher is the light accepting capability of the fiber.

3.2.4 *Fiber modes*

We have talked about the guidance of a light ray in optical fiber through NA and acceptance angle and have described that light incident between the normal and acceptance angle can propagate along the fiber core. But, it is not always the case, because the superposition of the incident travelling wave and reflected travelling wave (from the other end) results in constructive and destructive wave interferences that allow only certain modes or rays to propagate through optical fiber. Propagation of light through fiber can be explained rigorously by wave optics as light is an electromagnetic wave. But here we shall be brief about the fiber modes only. A mode can be defined as the distribution of an electromagnetic field in a transverse plane propagating through the optical fiber without changing its amplitude. The propagation of light into fiber, fiber mode, and fiber shape can completely be understood by knowing the all fiber parameters, i.e., core, cladding RIs (n_1, n_2), relative index difference (Δ), core radius (a), and wavelength of incident light (λ). These parameters are related altogether to define a characteristic wave guide parameter called V-number. It can be written as[58]:

$$V = \frac{2\pi a}{\lambda}\left(n_1^2 - n_2^2\right)^{1/2}, \tag{3.6}$$

When the V number is small, i.e., $V < 2.405$, only one mode is supported, while for V numbers greater than 2.405 the number of supported modes also increases. Moreover a fiber with parameter $V < 2.405$ is called single mode fiber and that with $V > 2.405$ is called multimode fiber. If $V > 10$, the fiber is a highly multimode fiber. In step index multimode fiber, number of modes (N) can be found by the following relation:

$$N \approx \frac{V^2}{2}, \tag{3.7}$$

While varieties of optical fibers exist according to fiber parameters, they have their own applications. For example, single mode fiber is highly in use for the communication purposes and for connecting different components in fiber optic devices. Moreover, multimode fibers are in use for sensing purposes, because they provide, robustness, strength, and a large surface area for interactions. Typical parameters of step index multimode silica fiber are $n_1 = 1.465$, $n_2 = 1.45$, $a = 25\,\mu\text{m}$, and $\Delta \approx 0.01$.

3.3 Optical Fiber Sensors

Optical fiber is mostly used in the field of communication but for the past few decades, they have got applications in the field of sensors due to their robustness, remote sensing, and online monitoring capabilities. Mostly multimode optical fibers are used for the chemical and biological sensing because they are robust and provide a large surface area for interaction, which further leads to increased dynamic sensing range of the analyte, and enhanced signal. Moreover, in the case of surface plasmons, multimode fibers provide easy and efficient excitation of surface plasmons by coupling of higher order (loosely bound) fiber modes to surface plasmons. In this section we shall discuss about the principle of optical fiber sensors, their designs and components in context to the components and transduction mechanisms discussed in earlier sections.

In principle, all kinds of receptors discussed earlier can be attached on the surface of an optical fiber. The kinds of receptor matrices which can be easily developed on metallic surfaces are favorable for SPR-based sensors. However, in case, if the receptor matrix does not fit compatible with metallic surfaces, such as the inhibition of urease by metals, the metallic surface can be coated with a thin layer of dielectric which can offer further binding of the receptor without loss of activity.[14]

Obviously, optical fiber based sensors are based on optical transduction. The optical fiber sensors are most generally the ones in which the sensor surface itself is generated on the fiber. In some cases, though, the fibers are integrated with other sensor surfaces just as waveguides for transporting the light from the source to the sensor and from there to the detector. In the first kind of sensors, the optical fibers themselves provide high quality substrate surface, as they themselves are drawn from the purest forms of glass. In such sensors, the cladding from small portion, in the middle, of an optical fiber is taken out, so that the evanescent field at the core-cladding interface is exposed to outside environment. This evanescent field interacts with the outer environment directly or through receptor and leads to a modulation in the guided light. The modulation in light is detected at the output end of the optical fiber by a suitable detector depending upon the mode of operation. For example, in sensors based on evanescent wave absorption, a change in intensity is measured by an optical power meter. While, in sensors employing spectral interrogation, the detection is carried out by a spectrometer. In some cases, the fiber optic circulators/couplers are employed and the transducer surface is developed on the end-face of the optical fiber. A rigorous discussion of different kinds of fiber geometries and sensors can be found elsewhere.[57] We limit here to the surface plasmon based transduction mechanism, which is again a part of optical transducers. Sensors based on surface plasmon scheme utilize most commonly the unclad fiber core surface for the coating of the metal film. They work in Kretschmann configuration of operation. The metal coated fiber core region acts as a transducer and converts any kind of analyte–receptor interactions into optical signals, which can be detected as a modulation of the incident light. The optical fibers having glass claddings are difficult to de-clad. Highly corrosive hydrofluoric acid (HF) is used for such purposes. Therefore, plastic clad silica (PCS) optical fibers are mostly used for such applications, where the cladding is made up of a plastic or polymer and the core is made of glass. Such a cladding can easily be removed by a sharp razor blade. Other advantage of PCS fibers over glass clad fibers is that we obtain a smooth fiber surface after un-cladding, while in glass clad fibers, uneven, non smooth surfaces are obtained after the HF etching. The method of fabrication of fiber optic plasmonic sensor probes has rigorously been discussed in Chapter 5.

In general, an optical fiber sensor assembly contains an optical source, fiber optic sensor probe, and a detector. The optical source could be a laser, light emitting diode (LED) or polychromatic light source such as tungsten halogen lamp, while the detector could be an optical power meter or a spectrometer, depending upon the mode of operation. The fiber optic sensor probe consists of the transducer and receptor in the sensing region to interact with the analyte. In the SPR-based optical fiber sensor devices, a polychromatic light source is used and a spectrometer works as a detector. Spectral interrogation mode of operation is carried out. The sensing phenomenon occurs in the metal coated sensor region, which is further coated with the receptors to specifically detect the analytes of interest.

3.4 Performance Parameters

The pros and cons of a sensor are evaluated in terms of the performance parameters. The faithfulness of any sensor depends on these parameters. We describe them as follows:

3.4.1 *Sensitivity*

Sensitivity of the sensor is defined as the measurable change in the output signal for unit change in the measurand of the analyte. In the case of fiber optic SPR sensors, the sensitivity is defined as the change in the resonance wavelength for a unit change in the refractive index, or the concentration of the analyte. High and linear sensitivity over the dynamical range of the sensor is always desirable but no sensor is ideal. Sensitivity also depends upon the environmental parameters and is quite critical to the method of fabrication as well. For SPR-based fiber optic refractive index sensor, the sensitivity is calculated as:

$$S = \frac{\Delta\lambda}{\Delta n} \quad \frac{\text{nm}}{\text{RIU}},$$

where $\Delta\lambda$ is the change in resonance wavelength (in nanometers) for a change of Δn in the refractive index of the sensing medium around the probe. RIU is the abbreviation for refractive index units.

3.4.2 *Selectivity/specificity*

Specificity/selectively of a sensor is termed as its capability to selectively determine the analyte concentration in a complex medium. The sensor must generate an original signal in the presence of a specific analyte and it should not create any false signal due to non specific binding of other foreign elements. For example, in the case of glucose sensor it should respond to glucose only and the signal shall not get interference from other molecules in the blood, i.e., plasma, cholesterol, triglycerides, etc. CO_2 gas sensor should not create false signals for other existing gases, i.e., oxygen, nitrogen, etc. Moreover, sensor could respond for a group of analytes where specificity is not an issue. For example, a phenolic compound sensor can detect the phenolic compounds in terms of concentration of these although sensitivity could not be the same for all phenolic compounds.[59] There is no quantitative measurement of specificity, but it is a property of the sensor, which can be checked by testing it for molecules similar to the analyte of interest and in real sensing environments.

3.4.3 *Limit of detection*

It is defined as the minimum detectable unit of the analyte by a sensor. SPR-based sensors have been found to have very low limits of detection.

3.4.4 *Accuracy*

How accurately a sensor can detect an analyte/sample or its concentration, is known as the accuracy. It also tells how much is the difference in a detected analyte concentration as compared to a standard and more accurate technique.

3.4.5 *Resolution*

It is defined as the capability of a sensor assembly to observe the smallest difference in the analyte. Actually the resolution is not the characteristic of the sensing probe but that of the detector. The greater the resolution of the detector is, the higher be the resolution of the sensor. In the case of spectroscopic sensors, the resolution of the sensor is defined by the resolution of the spectrometer and hence the resolution of a sensor is limited by that of the detector.

3.4.6 *Repeatability*

How many times a sensor can give similar results when put in the same analyte/environment is called the repeatability. It is desirable that all the sensors have good repeatability and long life.

3.4.7 *Reproducibility*

If a number of sensors are manufactured with the same principle for a particular analyte then the similarity of their response to the analyte is known as reproducibility. Different sensing probes of the same sensor shall give a reasonably good reproducibility. It does not matter whether the sensors are prepared at the same time or after the some period but sensors should reproduce the same results. Moreover, a sensor should give the same results whether experiments are performed after a long gap of time or at different places/laboratories by different operators. Repeatability and reproducibility are related terms; repeatability belongs to the process of taking measurements of one sensor for analyte in a short interval of time and seeking the same results.

3.4.8 *Noise*

It is defined as a random variation in the output signal of the sensor which carries no information. Noise can occur in an output signal without any input signal. It may be because of some environmental conditions and repetition of an analyte. Sensor has some noise itself in the device, called inherent noise. One could measure the inherent noise of the sensor by keeping it in a noise free environment and observing the output signal. For the good measurements, noise should always be minimum because it misleads the output signal.

3.4.9 *Range*

It is the range between the limit of detection till the maximum concentration which can be measured by the sensor. It is a characteristic of a sensor probe; however, the range of the sensor depends on the detecting device used to measure the output signal. Therefore, in some cases, the range of the sensor is limited by the range of the detector. Basically, sensors cannot work beyond their operating range. However, sometimes they can work beyond this range with extra and specific calibration conditions, (i.e., temperature, pressure). For example, a concentration sensor can increase its range of scan by application of a particular temperature. But the flaw is that if sensors are working beyond their range with specific calibration conditions, then they may produce an erratic signal. Further, suitable design of the sensor probe may lead to a change in the operating parameters and hence the range.

3.4.10 *Response time*

It is defined as the time taken by a sensor to generate a stable measurable output signal after adding the analyte on the receptor surface. Response time of a sensor could be from a few seconds to few days as well. A good sensor is supposed to have minimum response time. Optical fiber based sensors are very quick (low response time) and therefore can be used for real time monitoring.

3.4.11 *Linearity*

A good sensor is expected to have a linear calibration curve. It is linear when the output signal is directly proportional to analyte concentration throughout the range of the sensor, i.e., the slope of the curve of output vs. input signal is a constant.

3.4.12 *Drift*

Drift is a small and undesirable change in the characteristic curve of the sensor with time. Drift of a sensor decreases with time when its components become mature. It means that as the sensor becomes older, it loses its capacity to show any change in the response.

3.4.13 *Figure of merit*

Figure of merit (FOM) of a sensor is defined as the ratio of the sensitivity of the sensor and the width of the spectral curve. A good sensor always finds the minimum full width at half maximum (FWHM) of the spectrum and high sensitivity so that the FOM is high. The FOM is affected by few parameters such as concentration of the analyte, pH variation, temperature, etc.

3.5 Summary

In this chapter, we have discussed components and characteristics of a fiber optic sensor. The sensing mechanisms of various kinds used in sensors and their performance parameters have also been presented.

References

1. S.K. Srivastava, R. Verma and B.D. Gupta, Surface plasmon resonance based fiber optic sensor for the detection of low water content in ethanol, *Sensors and Actuators B* **153** (2011) 194–198.
2. S. Sen, V. Bhandarkar, K.P. Muthe, M. Roy, S.K. Deshpande, R.C. Aiyer, S.K. Gupta, J.V. Yakhmi and V.C. Sahni, Highly sensitive hydrogen sulphide sensors operable at room temperature, *Sensors and Actuators B* **115** (2006) 270–275.
3. G. Berruti, M. Consales, M. Giordano, L. Sansone, P. Petagna, S. Buontempo, G. Breglio and A. Cusano, Radiation hard humidity sensors for high energy physics applications using polyimide-coated fiber Bragg gratings sensors, *Sensors and Actuators B* **177** (2013) 94–102.
4. F. Qu, N.B. Li and H.Q. Luo, Highly sensitive fluorescent and colorimetric pH sensor based on polyethylenimine-capped silver nanoclusters, *Langmuir* **29** (2013) 1199–1205.
5. R. Verma and B.D. Gupta, Detection of heavy metal ions in contaminated water by surface plasmon resonance based optical fiber sensor using conducting polymer and chitosan, *Food Chemistry* **166** (2015) 568–575.
6. J. Kurittu, S. Lönnberg, M. Virta and M. Karp, A group-specific microbiological test for the detection of tetracycline residues in raw milk, *Journal of Agricultural and Food Chemistry* **48** (2000) 3372–3377.
7. N. Mitro, P.A. Mak, L. Vargas, C. Godio, E. Hampton, V. Molteni, A. Kreusch and E. Saez, The nuclear receptor LXR is a glucose sensor, *Nature* **445** (2007) 219–223.
8. A. Kaushik, P.R. Solanki, A.A. Ansari, G. Sumana, S. Ahmad and B.D. Malhotra, Iron oxide-chitosan nanobiocomposite for urea sensor, *Sensors and Actuators B* **138** (2009) 572–580.
9. A. Maria CF, B. Oliveira, M.H. Gil and A.P. Piedade, An electrochemical bienzyme membrane sensor for free cholesterol, *Journal of Electroanalytical Chemistry* **343** (1992) 105–115.
10. Y. Andreu, S. de Marcos, J.R. Castillo and J. Galbán, Sensor film for Vitamin C determination based on absorption properties of polyaniline, *Talanta* **65** (2005) 1045–1051.
11. I. Karube, Isao, Y. Wang, E. Tamiya and M. Kawarai, Microbial electrode sensor for vitamin B_{12}, *Analytica Chimica Acta* **199** (1987) 93–97.

12. R.O. Kadara, B.G.D. Haggett and B.J. Birch, Disposable sensor for measurement of vitamin B_2 in nutritional premix, cereal, and milk powder, *Journal of Agricultural and Food Chemistry* **54** (2006) 4921–4924.
13. W. Li, J. Chen, B. Xiang and D. An, Simultaneous on-line dissolution monitoring of multicomponent solid preparations containing vitamins B_1, B_{12} and B_6 by a fiber-optic sensor system, *Analytica Chimica Acta* **408** (2000) 39–47.
14. P. Bhatia and B.D. Gupta, Fabrication and characterization of a surface plasmon resonance based fiber optic urea sensor for biomedical applications, *Sensors and Actuators B* **161** (2012) 434–438.
15. S. Singh and B.D. Gupta, Fabrication and characterization of a surface plasmon resonance based fiber optic sensor using gel entrapment technique for the detection of low glucose concentration, *Sensors and Actuators B* **177** (2013) 589–595.
16. M. Boujtita, J.P. Hart and R. Pittson, Development of a disposable ethanol biosensor based on a chemically modified screen-printed electrode coated with alcohol oxidase for the analysis of beer, *Biosensors and Bioelectronics* **15** (2000) 257–263.
17. L. Pollegioni, L. Piubelli, S. Sacchi, M.S. Pilone and G. Molla, Physiological functions of D-amino acid oxidases: from yeast to humans, *Cellular and Molecular Life Sciences* **64** (2007) 1373–1394.
18. T. Vo-Dinh and B. Cullum, Biosensors and biochips: advances in biological and medical diagnostics, *Fresenius' Journal of Analytical Chemistry* **366** (2000) 540–551.
19. L.C. Clark and C. Lyons, Electrode systems for continuous monitoring in cardiovascular surgery, *Annals of the New York Academy of Sciences* **102** (1962) 29–45.
20. J.C. Huang, Y.F. Chang, K.H. Chen, L.C. Su, C.W. Lee, C.C. Chen, Yi-Ming Arthur Chen and C. Chou, Detection of severe acute respiratory syndrome (SARS) coronavirus nucleocapsid protein in human serum using a localized surface plasmon coupled fluorescence fiber-optic biosensor, *Biosensors and Bioelectronics* **25** (2009) 320–325.
21. A. Maines, M.I. Prodromidis, S.M. Tzouwara-Karayanni, M.I. Karayannis, D. Ashworth and P. Vadgama, An enzyme electrode for extended linearity citrate measurements based on modified polymeric membranes, *Electroanalysis*, **12** (2000) 1118–1123.
22. W.P. Yang, K. Green, S. Pinz-Sweeney, A.T. Briones, D.R. Burton and C.F. Barbas III, CDR walking mutagenesis for the affinity maturation of a potent human anti-HIV-1 antibody into the picomolar range, *Journal of Molecular Biology* **254** (1995) 392–403.
23. M.S. Wilson and N. Weiyan, Electrochemical multianalyte immunoassays using an array-based sensor, *Analytical Chemistry* **78** (2006) 2507–2513.
24. R. Polsky, J.C. Harper, D.R. Wheeler, S.M. Dirk, D.C. Arango and S.M. Brozik, Electrically addressable diazonium-functionalized antibodies for multianalyte electrochemical sensor applications, *Biosensors and Bioelectronics* **23** (2008) 757–764.

25. B. Johnsson, S. Löfås, G. Lindquist, Å. Edström, R.M. Müller Hillgren and A. Hansson, Comparison of methods for immobilization to carboxymethyl dextran sensor surfaces by analysis of the specific activity of monoclonal antibodies, *Journal of Molecular Recognition* **8** (1995) 125–131.
26. U. Bora, A. Sett and D. Singh, Nucleic acid based biosensors for clinical applications, *Biosensor J.* **1** (2003) 104.
27. E.A. Mothershed and A.M. Whitney, Nucleic acid based methods for the detection of bacterial pathogens: present and future considerations for the clinical laboratory, *Clinica Chimica Acta* **363** (2006) 206–220.
28. Y. Xiao, A.A. Lubin, A.J. Heeger and K.W. Plaxco, Label-free electronic detection of thrombin in blood serum by using an aptamer-based sensor, *Angewandte Chemie International Edition* **44** (2005) 5456–5459.
29. P.H. Lin, S.J. Tong, S.R. Louis, Y. Chang and W.Y. Chen, Thermodynamic basis of chiral recognition in a DNA aptamer, *Physical Chemistry Chemical Physics* **11** (2009) 9744–9750.
30. P. Corbisier, D. van der Lelie, B. Borremans, A. Provoost, Victor de Lorenzo, N.L. Brown, J.R. Lloyd, J.L. Hobman, E. CsoÈregid, G. Johanssond and B. Mattiasson, Whole cell- and protein-based biosensors for the detection of bioavailable heavy metals in environmental samples, *Analytica Chimica Acta* **387** (1999) 235–244.
31. D. Diamond, *Principles of chemical and biological sensors*, A Wiley-Interscience Publication, 1998.
32. P. Wang, Ping, G. Xu, L. Qin, Y. Xu, Y. Li and R. Li, Cell-based biosensors and its application in biomedicine, *Sensors and Actuators B: Chemical* **108** (2005) 576–584.
33. T. Wei, W. Tu, B. Zhao, Y. Lan, J. Bao, and Z. Dai, Electrochemical monitoring of an important biomarker and target protein: VEGFR2 in cell lysates, Nature Scientific Reports **4** (2014) 3982.
34. T.H. Rider, M.S. Petrovick, F.E. Nargi, J.D. Harper, E.D. Schwoebel, R.H. Mathews, D.J. Blanchard, L.T. Bortolin, A.M. Young, J. Chen and M.A. Hollis, A B cell-based sensor for rapid identification of pathogens, *Science* **301** (2003) 213–215.
35. S.E. Weigum, P.N. Floriano, N. Christodoulides and J.T. McDevitt, Cell-based sensor for analysis of EGFR biomarker expression in oral cancer, *Lab on a Chip* **7** (2007) 995–1003.
36. J.J. Pancrazio, J.P. Whelan, D.A. Borkholder, W. Ma and D.A. Stenger, Development and application of cell-based biosensors, *Annals of Biomedical Engineering* **27** (1999) 697–711.
37. S. Belkin and M.B. Gu, Whole Cell Sensing Systems I: reporter cells and devices. Springer, Heidelberg, 2010.
38. S.M. Tripathi, W.J. Bock, P. Mikulic, R. Chinnappan, A. Ng, M. Tolba and M, Long period grating based biosensor for the detection of Escherichia coli bacteria, *Biosensors and Bioelectronics* **35** (2012) 308–312.
39. J.S. Sidwell and G.A. Rechnitz, Bananatrode-An electrochemical biosensor for dompamine, *Biotechnology Letters* **7** (1985) 419–422.

40. C.A. Sanders, M. Rodriguez Jr and E. Greenbaum, Stand-off tissue-based biosensors for the detection of chemical warfare agents using photosynthetic fluorescence induction, *Biosensors and Bioelectronics* **16** (2001) 439–446.
41. N.J. Ronkainen, H.B. Halsall and W.R. Heineman, Electrochemical biosensors, *Chemical Society Review* **39** (2010) 1747–1763.
42. A.L. Ghindilis, P. Atanasov, M. Wilkins and E. Wilkins, Immunosensors: electrochemical sensing and other engineering approaches, *Biosensors and Bioelectronics* **13** (1998) 113–131.
43. Z. Muhammad-Tahir and E.C. Alocilja, A conductometric biosensor for biosecurity, *Biosensors and Bioelectronics* **18** (2003) 813–819.
44. J. Wang, Analytical electrochemistry, John Wiley & Sons, 2006.
45. C. K. O'Sullivan, and G. G. Guilbault, Commercial quartz crystal microbalances — theory and applications, Biosensors and Bioelectronics **14** (1999) 663–670.
46. B. Serra, M. Gamella, A.J. Reviejo and J.M. Pingarron, Lectin-modified piezoelectric biosensors for bacteria recognition and quantification, *Analytical and Bioanalytical Chemistry* **391** (2008) 1853–1860.
47. P. Skládal, C.D.S. Riccardi, H. Yamanaka and P.I. da Costa, Piezoelectric biosensors for real-time monitoring of hybridization and detection of hepatitis C virus, *Journal of Virological Methods* **117** (2004) 145–151.
48. C. Yao, T. Zhu, J. Tang, R. Wu, Q. Chen, M. Chen, B. Zhang, J. Huang and W. Fu, Hybridization assay of hepatitis B virus by QCM peptide nucleic acid biosensor. *Biosensors and Bioelectronics* **23** (2008) 879–885.
49. B. Xie, K. Ramanathan and B. Danielsson, Principles of enzyme thermistor systems: Applications to biomedical and other measurements. In *Thermal Biosensors, Bioactivity, Bioaffinitty*, Berlin, Heidelberg: Springer. 1999 pp. 1–33.
50. M.L. Antonelli, C. Spadaro and R.F. Tornelli, A microcalorimetric sensor for food and cosmetic analyses: l-Malic acid determination, *Talanta* **74** (2008) 1450–1454.
51. T. Maskow, K. Wolf, W. Kunze, S. Enders and H. Harms, Rapid analysis of bacterial contamination of tap water using isothermal calorimetry, *Thermochimica Acta* **543** (2012) 273–280.
52. P. Kirchner, J. Oberländer, P. Friedrich, J. Berger, G. Rysstad, M. Keusgen and M.J. Schöning, Realisation of a alorimetric gas sensor on polyimide foil for applications in aseptic food industry, *Sensors and Actuators B* **170** (2012) 60–66.
53. P. Bataillard, Calorimetric sensing in bioanalytical chemistry: Principles, applications and trends, *TrAC Trends in Analytical Chemistry* **12** (1993) 387–394.
54. I. Biran, X. Yu and D.R. Walt, Optrode-based fiber optic biosensors (bio-optrode). In *Optical Biosensors: Today and Tomorrow*, F. Ligler, & C. Taitt (Eds.) Amsterdam: Elsevier. 2008, pp. 3–82.
55. D. Atias, Y. Liebes, V. Chalifa-Caspi, L. Bremand, L. Lobel, R.S. Marks and P. Dussart, Chemiluminescent optical fiber immunosensor for the detection

of IgM antibody to dengue virus in humans, *Sensors and Actuators B* **140** (2009) 206–215.

56. K. Thyagarajan and A. Ghatak, *Fiber Optic Essentials*, John Wiley & Sons, 2007.
57. B.D. Gupta, *Fiber Optic Sensors: Principles and Applications*, New India Publishing Agency, New Delhi, 2006.
58. A. Ghatak and K. Thyagarajan, *An Introduction to Fiber Optics*, Cambridge University Press, 1998.
59. S. Singh, S.K. Mishra and B.D. Gupta, SPR based fibre optic biosensor for phenolic compounds using immobilization of tyrosinase in polyacrylamide gel, *Sensors and Actuators B* **186** (2013) 388–395.

Chapter 4

Theory of SPR-based Optical Fiber Sensor

4.1 Introduction

In Chapter 2, we have introduced propagating surface plasmons. These are collective oscillations of free electrons at the metal–dielectric interface resulting in a TM-polarized wave propagating along the interface with the maximum field at the interface decaying exponentially in both the media. The propagating surface plasmons can be excited by p-polarized light and the resonance or the maximum transfer of energy from the excitation light to the surface plasmons occurs when the propagation vectors of the two become equal. To excite surface plasmons Kretschmann configuration, discussed in Chapter 2, is used where the base of a high refractive index prism is coated with a thin metal layer and the dielectric medium is kept on the other side of the metal layer. For the excitation of surface plasmons, light is incident on the prism–metal interface through one of the faces of the prism. When the angle of incidence is greater than the critical angle the evanescent wave of the totally internally reflected light excites the surface plasmons at the metal–dielectric interface. The same principle can be applied when an optical fiber is used in place of the prism for the excitation of the surface plasmons and hence for the sensing of the refractive index of the dielectric medium. In an optical fiber, light propagates through the phenomenon of total internal reflection (TIR) and hence the evanescent wave of the propagating ray can be used to excite the

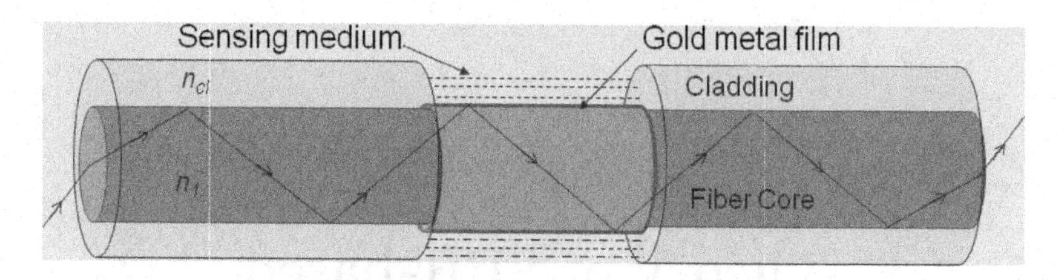

Fig. 4.1. Schematic of the surface plasmon resonance based optical fiber sensing probe. Reprinted with permission from ref. [1].

surface plasmons, if the metal layer is coated over the unclad core of the fiber. A schematic of the optical fiber based propagating surface plasmon resonance (SPR) sensing probe is shown in Fig. 4.1.[1] The sensing region of a fiber optic SPR sensor, in general, comprises of the fiber core, a thin metal layer, and a dielectric sensing medium with a refractive index lower than the refractive index of the fiber core. In the case of a fiber optic SPR sensor with bimetallic layers, the single metal layer is replaced by two adjacent layers of two different metals. In the case of guided wave SPR sensors, discussed in Sec. 2.6, the metal layer is covered with a very thin layer of high refractive index dielectric material, which is further surrounded by the sensing medium. In general, for the bio-sensing applications, the fiber optic sensor configurations involve further immobilization of biomolecular recognition elements (BREs) on the sensor surface, which come in contact with the sensing medium comprising of analyte of interest.

Generally, step-index multi-mode plastic clad silica (PCS) optical fiber is used for the fabrication of SPR sensors. Since the fiber optic SPR sensors employ wavelength/spectral interrogation, as discussed in Chapter 2, the wavelength dependencies of the dielectric constants of both the fiber core and the metal are required for the purpose of modeling and simulation of the sensor. The wavelength dependence of the refractive index (n_1) of a silica core optical fiber is given by the following Sellmeier dispersion relation:

$$n_1(\lambda) = \sqrt{1 + \frac{a_1\lambda^2}{\lambda^2 - b_1^2} + \frac{a_2\lambda^2}{\lambda^2 - b_2^2} + \frac{a_3\lambda^2}{\lambda^2 - b_3^2}}, \tag{4.1}$$

where a_1, a_2, a_3, b_1, b_2, and b_3 are the Sellmeier coefficients and λ is the wavelength (in μm) of light propagating in the medium.[2] The values of the coefficients are given in Appendix B. The dispersion relation of the metal layer is given by the Drude model[3] discussed in Appendix A, according to which the dielectric constant of a metal is given by:

$$\varepsilon_m(\lambda) = \varepsilon_{mr} + i\varepsilon_{mi} = 1 - \frac{\lambda^2 \lambda_c}{\lambda_p^2 (\lambda_c + i\lambda)}, \tag{4.2}$$

where λ_p and λ_c are the plasma and collision wavelengths, respectively. The values of these characteristic wavelengths for various metals are given in Appendix B. In the case of multiple layers of different materials on the fiber core, the dispersion relations for all these layers are required for simulations. In fiber optic SPR sensor, unpolarized light is launched into the fiber although the TM polarized light excites the surface plasmons. This is because the geometry of the fiber is cylindrical and it is not possible to maintain the polarization of the incident light. Therefore, rather talking in terms of transverse electric (TE) and transverse magnetic (TM) polarized light, we simply say that only the component of E-vector of light which is perpendicular to the core-metal interface excites the surface plasmons at the metal-sensing medium interface. As discussed in earlier chapters, the light guided in the optical fiber core excites the surface plasmons at the metal and sensing medium interface with the following resonance condition: matching of frequencies and wave-vectors of the excitation wave and that of surface plasmon wave. At the resonance condition, the maximum incident energy is transferred to the surface plasmons and a sharp dip is obtained in the spectrum of the transmitted output power. The wavelength corresponding to the minimum transmitted power is called resonance wavelength as defined in Chapter 2. Since an unpolarized light is used for the excitation of surface plasmons in the fiber optic SPR sensor, the transmitted power at the output end of the fiber will be the sum of the affected perpendicular polarization (TM-polarized) corresponding to SPR and the unaffected polarization (TE-polarized). The transmission spectrum, therefore, suffers with less contrast which comes due to a DC background of about 50% of TE polarized light. The presence of other polarizations, however, does not affect the position of the dip in SPR spectrum or the resonance wavelength. Moreover, the resonance wavelength is highly dependent on the dielectric properties of metal layer and sensing medium. Different resonance wavelengths correspond to different metals used for generating the surface plasmons and resonance wavelength changes with the change in the refractive index of the sensing medium or sensing surface (in the case of biosensors). For simulations, the highly multi-mode optical fiber of large core diameter is used which can be approximated by a planar waveguide approach with TM-polarized light. To know the spectrum of the transmitted power, the amplitude reflection coefficient at the interface is required. This is determined by using N-layer model. Once the amplitude reflection coefficient is known, the transmitted power can be calculated with the help of the power distribution of light guided in the core of an optical fiber, which depends on the light launching condition and the light source.

4.2 N-Layer Model

In general, the expression for amplitude reflection coefficient of a multi-layer structure for a p-polarized incident beam is determined using transfer matrix method for N-layer model.[4,5] In this method, N numbers of layers (or, $N - 1$ numbers of interfaces) of different optical materials are assumed to be stacked along the direction of propagation of light (see Fig. 4.2). The arbitrary medium layer (k^{th} layer) in the figure is designated by thickness d_k, dielectric constant ε_k, permeability μ_k, and refractive index n_k.

In a SPR-based conventional sensor the minimum number of layers is three and hence first we consider three layers geometry in which a dielectric layer of refractive index n_2 is sandwiched between two semitransparent media of refractive indices n_1 and n_3, as shown in Fig. 4.3.

Consider a linearly polarized wave impinging on the dielectric film and both the electric ($\boldsymbol{U}$) and magnetic ($\boldsymbol{V} = \boldsymbol{B}/\mu$) fields

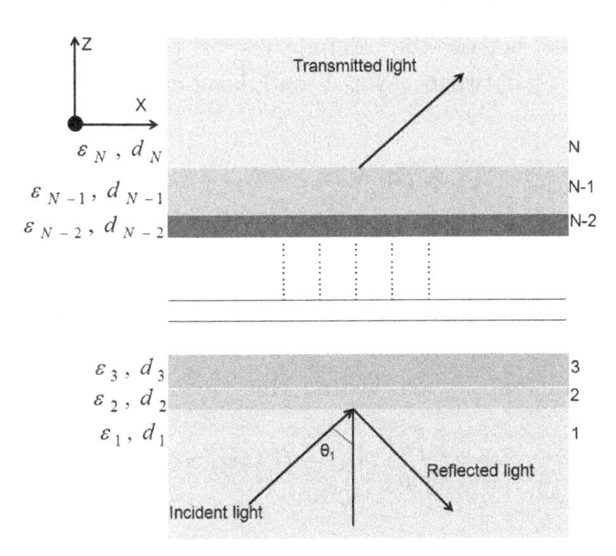

Fig. 4.2. Schematic of a stack of N-layers of different optical materials and different thicknesses.

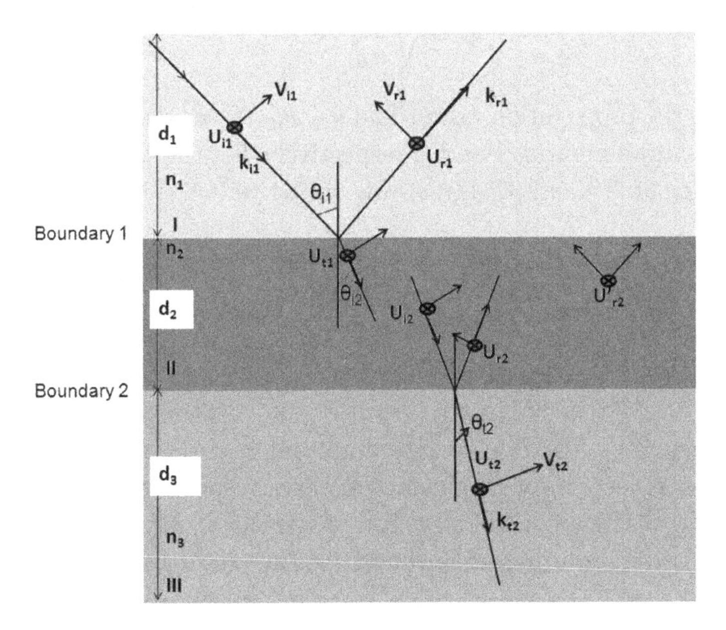

Fig. 4.3. Three layer system and fields at the boundaries.

are continuous across the boundaries or the interfaces. At the first boundary (between layer 1 and layer 2) the electric field is given as:

$$U_1 = U_{i1} + U_{r1} = U_{t1} + U'_{r2}, \tag{4.3}$$

where U_{i1}, U_{r1}, and U_{t1} are the amplitudes of electric fields of incident, reflected and transmitted wave at boundary 1, whereas U'_{r2} is the amplitude of the electric field of the reflected wave from boundary 2 and incident on boundary 1. Applying boundary condition for magnetic field at boundary 1, one obtains:

$$\begin{aligned} V_1 &= \sqrt{\frac{\varepsilon_1}{\mu_1}}\left(U_{i1} - U_{r1}\right) n_1 \cos\theta_{i1} \\ &= \sqrt{\frac{\varepsilon_1}{\mu_1}}\left(U_{t1} - U'_{r2}\right) n_2 \cos\theta_{i2}, \end{aligned} \tag{4.4}$$

where U and V are related as:

$$V = \sqrt{\frac{\varepsilon_1}{\mu_1}} n_1 \hat{k} \times U, \tag{4.5}$$

$\hat{k}$ being the propagation vector and θ_{i1}, θ_{i2} the angles of incidence of the rays at boundaries 1 and 2, respectively. For the second boundary (interface of 2nd and 3rd layer), we can write:

$$U_2 = U_{i2} + U_{r2} = U_{t2}, \tag{4.6}$$

and

$$V_2 = \sqrt{\frac{\varepsilon_1}{\mu_1}}\left(U_{i2} - U_{r2}\right) n_2 \cos\theta_{i2} = \sqrt{\frac{\varepsilon_1}{\mu_1}} U_{t2} n_3 \cos\theta_{t2}, \tag{4.7}$$

where U_{i2}, U_{r2} and U_{t2} are the amplitudes of the electric fields of incident, reflected, and transmitted waves respectively at boundary 2 and θ_{t2} is the angle of refraction of the ray at boundary 2. When a wave traverses a film of thickness d_2, the phase shift occurred can be written as:

$$\beta_2 = \frac{2\pi d_2}{\lambda}\left(\varepsilon_2 - n_1^2 \sin^2\theta_1\right)^{1/2}, \tag{4.8}$$

The field equations at boundary 2 can then be written as:

$$U_{i2} = U_{t1}e^{-i\beta_2}, \tag{4.9}$$

$$U_{r2} = U'_{r2}e^{i\beta_2}, \tag{4.10}$$

Hence, Eqs. (4.6) and (4.7) can be written as:

$$U_2 = U_{t1}e^{-i\beta_2} + U'_{r2}e^{+i\beta_2}, \tag{4.11}$$

$$V_2 = \sqrt{\frac{\varepsilon_1}{\mu_1}}\left(U_{t1}e^{-i\beta_2} - U'_{r2}e^{+i\beta_2}\right) n_2 \cos\theta_{i2}, \tag{4.12}$$

After solving Eqs. (4.11) and (4.12) for U_{t1} and U'_{r2} and substituting their values in Eqs. (4.3) and (4.4) one obtains:

$$U_1 = U_2 \cos\beta_2 - V_2(i \sin\beta_2)/q_2, \tag{4.13}$$

$$V_1 = U_2 q_2(-i \sin\beta_2) + V_2(\cos\beta_2), \tag{4.14}$$

where

$$q_2 = \frac{\left(\varepsilon_2 - n_1^2 \sin^2\theta_{i1}\right)^{1/2}}{\varepsilon_2},$$

In matrix notation, Eqs. (4.13) and (4.14) together can be written as:

$$\begin{bmatrix} U_1 \\ V_1 \end{bmatrix} = \begin{bmatrix} \cos\beta_2 & (-i\sin\beta_2)/q_2 \\ -iq_2\sin\beta_2 & \cos\beta_2 \end{bmatrix} \begin{bmatrix} U_2 \\ V_2 \end{bmatrix}, \tag{4.15}$$

or

$$\begin{bmatrix} U_1 \\ V_1 \end{bmatrix} = m_2 \begin{bmatrix} U_2 \\ V_2 \end{bmatrix}, \tag{4.16}$$

where

$$m_2 = \begin{bmatrix} \cos\beta_2 & (-i\sin\beta_2)/q_2 \\ -iq_2\sin\beta_2 & \cos\beta_2 \end{bmatrix}, \tag{4.17}$$

In generalized form, electric and magnetic fields at the first boundary $Z = Z_1$ are related to those at the final boundary $Z = Z_{N-1}$ by:

$$\begin{bmatrix} U_1 \\ V_1 \end{bmatrix} = M \begin{bmatrix} U_{N-1} \\ V_{N-1} \end{bmatrix}, \tag{4.18}$$

where U_1 and V_1, respectively, are the tangential components of electric and magnetic fields at the boundary of first and second layers, while U_{N-1} and V_{N-1} are the corresponding fields at the boundary of $(N-1)^{\text{th}}$ and N^{th} layers. The characteristic matrix of the combined structure, M is given by:

$$M = \prod_{k=2}^{N-1} m_k = \begin{bmatrix} M_{11} & M_{12} \\ M_{21} & M_{22} \end{bmatrix}, \tag{4.19}$$

where k is the number of layers amongst 1st and N^{th} layer. There are $N-1$ interfaces for which we can find the characteristic matrices from the following general matrix:

$$m_k = \begin{bmatrix} \cos\beta_k & (-i\sin\beta_k)/q_k \\ -iq_k\sin\beta_k & \cos\beta_k \end{bmatrix}, \tag{4.20}$$

where

$$q_k = \left(\frac{\mu_k}{\varepsilon_k}\right)^{1/2}\cos\theta_k = \frac{\left(\varepsilon_k - n_1^2\sin^2\theta_1\right)^{1/2}}{\varepsilon_k}, \tag{4.21}$$

and

$$\beta_k = \frac{2\pi}{\lambda} n_k \cos\theta_k \left(z_k - z_{k-1}\right) = \frac{2\pi\, d_k}{\lambda}\left(\varepsilon_k - n_1^2\sin^2\theta_1\right)^{1/2}, \tag{4.22}$$

In present notations, the refractive index of a medium, n, is related to the dielectric constant, ε, as $n^2 = \varepsilon$ and $\theta_1 = \theta_{i1}$.

If there are three layers (fiber core, metallic layer, and sensing medium) then $N = 3$. Therefore, accordingly, n_1, n_2 and n_3 can be assigned as the refractive indices of the core of the optical fiber, metal layer, and the sensing medium, respectively. The transfer matrices m_k are calculated as per the earlier equations for all the interfaces and then multiplied to obtain the final transfer matrix M of the multi-layer structure. After simplification and rearrangement of the algebraic expressions, the expression for the amplitude reflection coefficient (r_p) for p-polarized wave is given by:

$$r_p = \frac{\left(M_{11} + M_{12}q_3\right)q_1 - \left(M_{21} + M_{22}q_3\right)}{\left(M_{11} + M_{12}q_3\right)q_1 + \left(M_{21} + M_{22}q_3\right)} \tag{4.23}$$

In the Eq. (4.23), q_1 corresponds to fiber core and q_3 to the sensing medium. Both these terms are calculated from Eq. (4.21) by putting the refractive index, (i.e., n_1 and n_3) or dielectric constants, (i.e., ε_1 and ε_3) and the value of the angle of incidence and M_{ij} $(i, j = 1, 2)$ is one of the components of the transfer matrix M, given by Eq. (4.19). Finally, the reflectance (R_p) for p-polarized light as a function of angle of incidence is given by:

$$R_p = |r_p|^2 \tag{4.24}$$

The N-layer model can be extended to any number of layers. Also, the calculations are very easy, accurate, and precise due to the absence of approximations. To determine the effective transmitted power, the reflectance (R_p) for a single reflection is raised to the power of the number of reflections a specific propagating ray undergoes at the sensor interface. The number of reflections depends on the angle of the incident ray and the light launching condition. In the following sections, we calculate the effective transmitted power off the output end of the optical fiber for different light launching conditions and source profiles. In fiber optic SPR sensors, generally spectral interrogation is employed because all the guided modes are launched. We first consider a simple geometry with on axis excitation, in which the light from a collimated broadband source is focused on the axial point of the input end face of the optical fiber. Later, we shall discuss the excitation with off-axis rays and their effect. In the last, we shall discuss the case of diffuse source such as LED (light emitting diode).

The power profile of light source plays a very important role in the efficient excitation of guided modes in an optical fiber and determines the power profile of the guided mode. The distribution of power in all the rays emitted from each differential surface of the source is defined as its power profile. To illustrate the concept of power profile of a source, we consider a small element of the light source with surface area dS as shown in Fig. 4.4. Let us consider that each point of the source emits light in a cone of ψ half angle. In that case, the power emitted in a cone of solid angle $d\Omega$ at an angle of ψ_Ω from the normal to the area element dS is

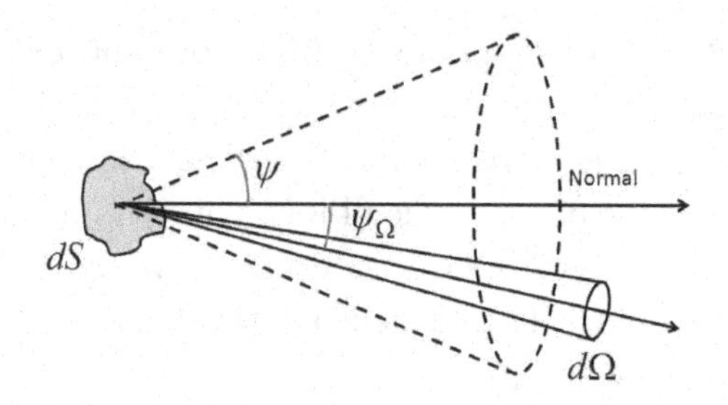

Fig. 4.4. Illustration of illumination from a small area of a source.

given by[6]:

$$dP = \begin{Bmatrix} I(\psi_\Omega)dSd\Omega, & for\ 0 \le \psi_\Omega \le \psi; \\ 0, & for\ \psi \le \psi_\Omega \le \pi/2 \end{Bmatrix}, \tag{4.25}$$

where $I(\psi_\Omega)$ is the intensity at an angle ψ_Ω. Based on the earlier discussion, we may now define the collimated and diffuse sources. A source is called collimated for which the angle ψ is zero. To say, each differential element dS of the source radiates along its axial direction only. For a diffuse source, the angle ψ becomes equal to $\pi/2$. For such a source, each differential element of the source emits in all the directions, at angles ranging from 0 to $\pi/2$. For a better understanding, we present next the schematic diagrams for both collimated and diffuse sources in Fig. 4.5. We, now, discuss one by one the cases of on-axis and off-axis excitation of guided rays in an optical fiber when light from a collimated source is focused on its end face.

4.3 Excitation by Meridional Rays: On Axis Excitation

For calculating the normalized transmitted power using N-layer model, we assume that the optical fiber is unclad from the middle and then coated with a SPR active metal layer, which is further surrounded by a dielectric sensing medium. The light from a broadband source is first collimated and then focused onto one of the ends and at the axial point of the optical fiber with the help of a microscope objective as shown in Fig. 4.6. The light transmitted through the fiber is collected at the other end by a spectrometer for

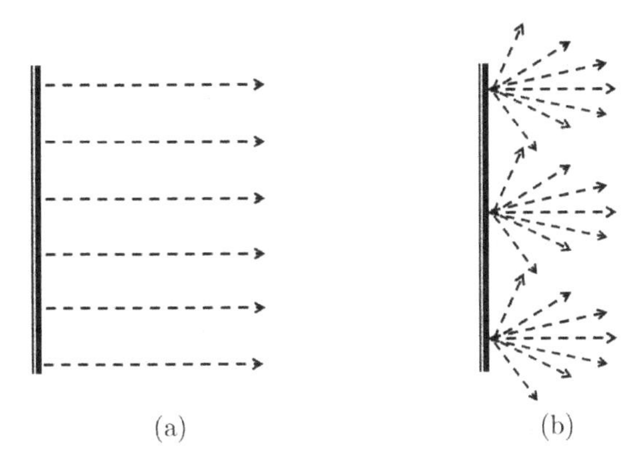

Fig. 4.5. Schematic ray diagram for (a) collimated source, and (b) diffuse source.

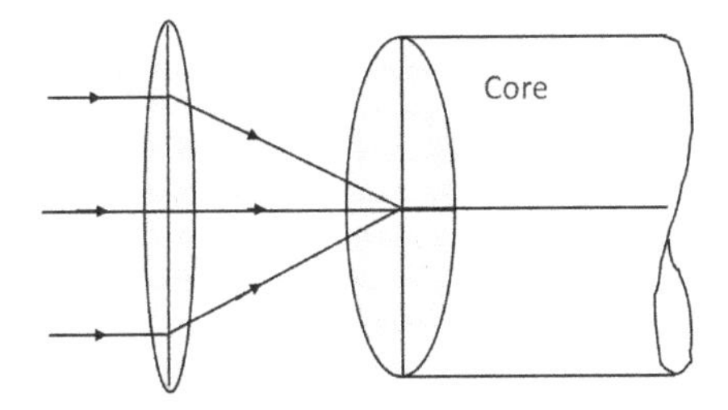

Fig. 4.6. Excitation of meridional rays in an optical fiber.

recording its spectrum. For simplicity we have assumed a zero spot size of the focused beam on the fiber end face.

To calculate the optical power transmitted through the optical fiber, all the guided rays must be considered in the simulations. The power profile of the guided rays in an optical fiber is specified by the distribution of optical power among all the guided ray directions, which depends basically on the light launching condition. In the present section, we discuss the excitation of surface plasmon by meridional rays in an optical fiber. Meridional rays are those kinds of rays which originate from the off-axis point and lie in the meridian plane; meridian plane contains off-axis point object and the optical axis and constitutes a plane of symmetry for the whole system.

In Fig. 4.7, we show launching of two meridional rays and their guidance inside the fiber core. The optical power, dP, arriving at

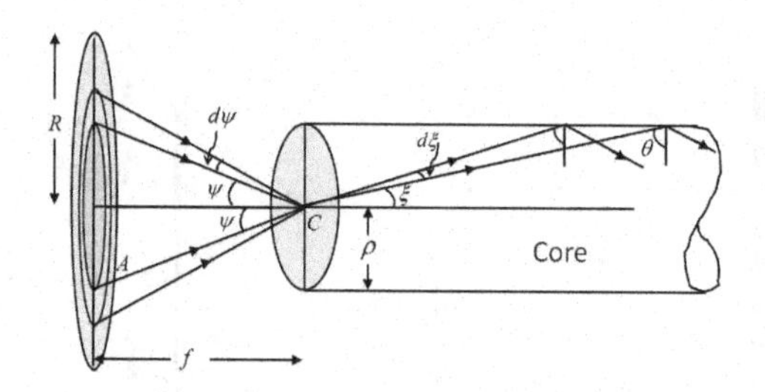

Fig. 4.7. Launching of two meridional rays with angles ψ and $\psi + d\psi$ with the axis of the fiber.

point C due to the rays confined between the angles ψ and $\psi + d\psi$ is given by[5]:

$$dP = 2\pi f^2(\tan\psi/\cos^2\psi)d\psi, \tag{4.26}$$

where f is the focal length of the lens focusing the collimated light beam on the face and axis of the optical fiber.

Since, the range of angles for the rays guided into the optical fiber due TIR varies from the critical angle of the core-cladding interface (θ_{cr}) to $\pi/2$, it is more convenient to calculate the power in terms of θ, rather than ψ, because for the modeling of SPR sensor using N-layer model the angles used are with respect to the normal to the fiber core-metal interface. According to Snell's law, for a ray refracting at the interface of two media with refractive indices n_1 and n_2:

$$n_1 \sin\theta_1 = n_2 \sin\theta_2, \tag{4.27}$$

where θ_1 and θ_2 are the angles of the ray with the normal to the interface in corresponding media. For a ray incident on the optical fiber end face at an angle ψ from air ($n = 1$) and refracted at an angle ξ (see Fig. 4.7), Eq. (4.27) can be rewritten as:

$$\sin\psi = n_1 \sin\xi, \tag{4.28}$$

where n_1 is the refractive index of the core of the optical fiber. The angles θ and ψ are related as:

$$\left.\begin{aligned} \cos^2\psi &= (1 - n_1^2\cos^2\theta), \\ \tan\psi &= \frac{n_1\cos\theta}{\sqrt{1 - n_1^2\cos^2\theta}} \end{aligned}\right\}, \tag{4.29}$$

Use of Eq. (4.29) in Eq. (4.26) gives:

$$dP = \frac{-2\pi f^2 n_1^2 \sin\theta\cos\theta}{\left(1 - n_1^2\cos^2\theta\right)^2} d\theta, \tag{4.30}$$

Thus the angular power distribution inside the optical fiber due to a guided ray incident at an angle θ with respect to the normal to the core-cladding interface, irrespective of the collimating lens, is given by:

$$dP \propto \frac{n_1^2 \sin\theta\cos\theta}{\left(1 - n_1^2\cos^2\theta\right)^2} d\theta, \tag{4.31}$$

Therefore, for p-polarized light, considering all the guided rays, the generalized expression for the normalized transmitted power, P_{trans}, for a fiber optic SPR sensor can be given as:

$$P_{\text{trans}} = \frac{\int_{\theta_{\text{cr}}}^{\pi/2} R_p^{N_{\text{ref}}(\theta)} \frac{n_1^2 \sin\theta\cos\theta}{\left(1-n_1^2\cos^2\theta\right)^2} d\theta}{\int_{\theta_{\text{cr}}}^{\pi/2} \frac{n_1^2 \sin\theta\cos\theta}{\left(1-n_1^2\cos^2\theta\right)^2} d\theta}, \tag{4.32}$$

where R_p is the reflectance for the ray with polarization normal to the core-cladding interface and is defined in Eq. (4.24) for N-layer model; $N_{\text{ref}}(\theta)$ represents the total number of reflections performed by a ray making an angle θ with the normal to the core-metal layer interface in the sensing region, while the critical angle of the core-cladding interface for a given optical fiber is given by:

$$\theta_{\text{cr}} = \sin^{-1}\left(\frac{n_{cl}}{n_1}\right), \tag{4.33}$$

where n_{cl} is the refractive index of the fiber cladding and:

$$N_{\text{ref}}(\theta) = \frac{L}{2\rho\tan\theta}, \tag{4.34}$$

L and ρ represent the length of the exposed sensing region and the fiber core radius, respectively. The output transmitted power, P_{trans}, is calculated from Eq. (4.32) for the case of excitation of surface plasmons using on-axis light launching in the fiber. The on-axis excitation of surface plasmons is taken as the case of conventional

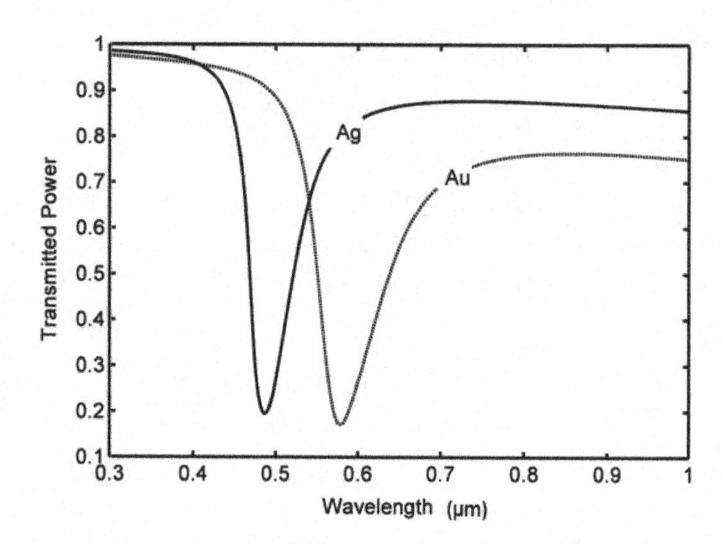

Fig. 4.8. SPR curves for fiber optic SPR sensors with layer of two different metals (silver and gold) over unclad core of the fiber for 1.333 refractive index of the sensing medium.

SPR. The SPR curves determined for gold and silver metal layers using N-layer model and Eq. (4.32) are shown in Fig. 4.8.

For simulation, following values of the parameters of the optical fiber have been used: core radius $\rho = 300\,\mu\text{m}$, numerical aperture (NA) $= 0.24$, sensing length L $= 1$ cm and core refractive index from Sellmeier relation given by Eq. (4.1), with the values of constants given in Appendix B. The dispersion relation for silver and gold are used from Drude model with the collision and plasma wavelengths given in Appendix B for both the metals. The thicknesses of silver and gold layers used are 40 nm and 50 nm respectively and the sensing medium is water with refractive index 1.333.

The SPR wavelength which corresponds to the minimum transmitted power is determined from the curve plotted between the transmitted output power and the wavelength. In the case of gold and silver, the minimum transmitted power is obtained at different wavelengths as shown in Fig. 4.8. For the fiber and the probe parameters used for the simulation, the resonance wavelengths for

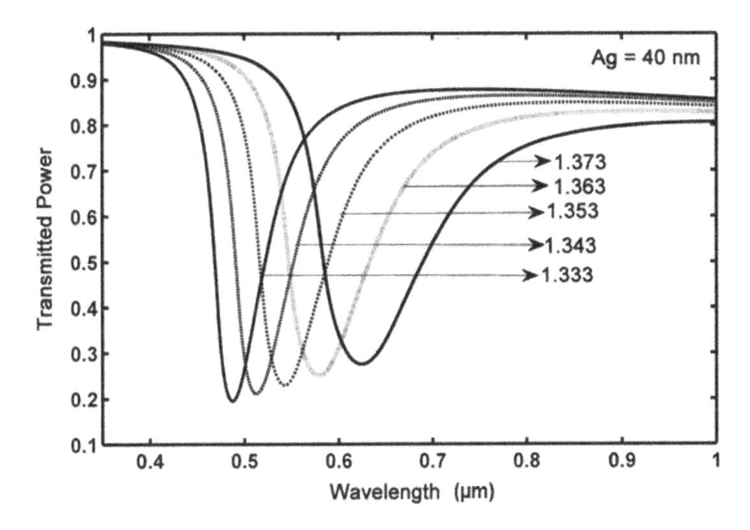

Fig. 4.9. SPR spectra for different refractive indices of the sensing medium for a silver coated optical fiber SPR sensor.

silver and gold coatings are 480.2 nm and 578.7 nm, respectively. The resonance wavelength is different for different plasmonic metals because of specific collision and plasma wavelengths for each metal. Due to the same reason, the width of the SPR curve is different for different metals.

The SPR spectra for thin layer of silver metal coated optical fiber sensor for different refractive indices of the sensing medium are shown in Fig. 4.9. The parameters used for simulation are the same as in the case of Fig. 4.8 for silver. It can be seen that the resonance wavelength is different for different refractive indices of the sensing medium. As the refractive index increases, the SPR curve shifts towards higher wavelengths. In other words, the resonance wavelength increases with the increase in the refractive index of the sensing medium. Figure 4.10 shows the variation of resonance wavelength with the refractive index of the sensing medium for the SPR spectra plotted in Fig. 4.9. It may also be seen from Fig. 4.9 that as the refractive index increases the broadening of SPR curves increases which implies that the detection accuracy of the sensor decreases.

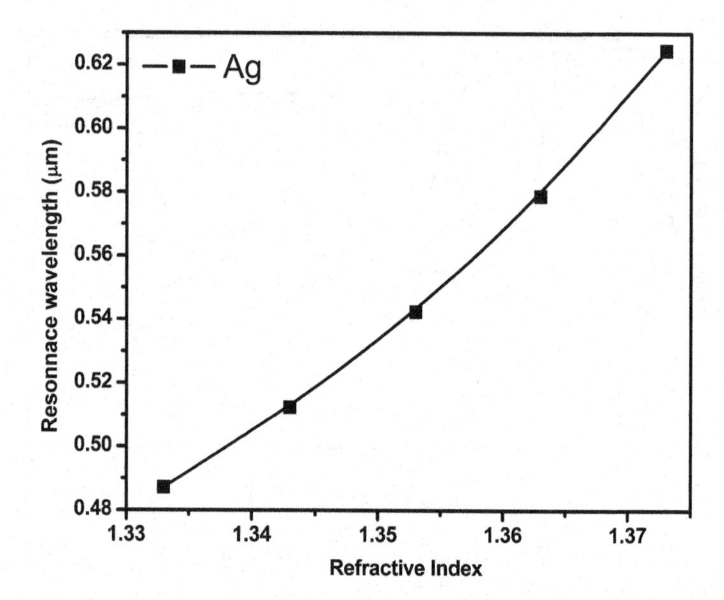

Fig. 4.10. Variation of resonance wavelength with the refractive index of the sensing medium for silver coated SPR-based fiber optic refractive index sensor.

4.4 Excitation by Skew Rays: Off Axis Excitation

In Sec. 4.3, we have considered the excitation of surface plasmons by meridional rays only and for that we have used the collimated beam and focusing element arrangement in such a way that light is focused at the axial point on the end face of the fiber as shown in Fig. 4.6. As shown in Fig. 4.11, if the collimated beam is focused at a point other than the axial point on the end face of the optical fiber, then both the meridional and skew rays will get excited in the fiber core.[7] As mentioned earlier, meridional rays are those rays which intersect the fiber axis between reflections, while the skew rays never intersect the fiber axis.

Before we proceed further, we shall discuss briefly the excitation and properties of skew rays. The skew rays are the guided rays in the optical fiber which do not intersect the axis of the fiber. Such rays travel on the boundary of the optical fiber and follow a nearly helical path inside the fiber (see Figs. 4.12(a) and (b)). To specify the

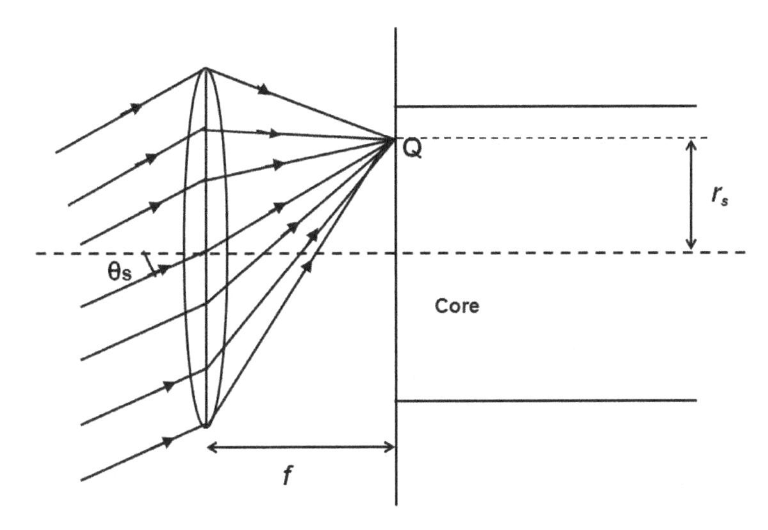

Fig. 4.11. Off-axis excitation: in this kind of light launching both meridional and skew rays are guided in the fiber.

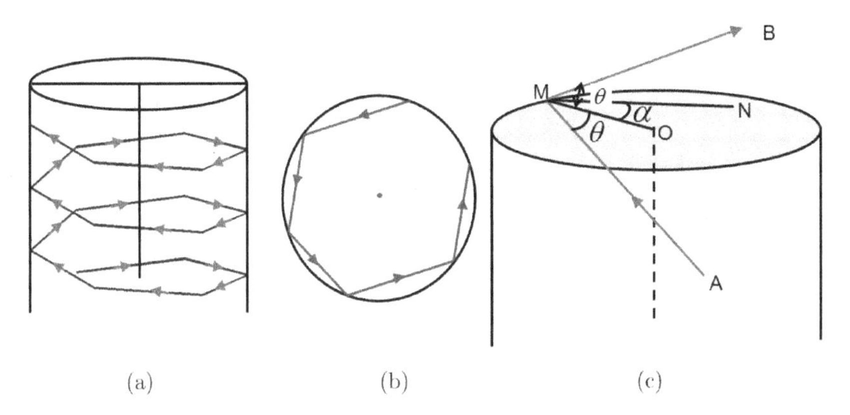

Fig. 4.12. Skew rays in optical fiber: (a) Schematic of the optical path of skew rays in an optical fiber, (b) Skew rays in transverse cross-section of the optical fiber, and (c) Schematic representation of skewness angle.

trajectory of a skew ray, a second angle known as skewness angle (α) is defined. As shown in Fig. 4.12(c), a ray AM incident at an angle θ from the normal to the core-cladding interface (OM) gets total internally reflected at an angle θ in the direction MB. The angle between the projections of the total internally reflected ray MB on the cross-section of the fiber core (MN) and the normal to the core-cladding interface (OM) is termed as skewness angle.

The midpoint of the skew rays between successive reflections touch a cylindrical surface of radius r_s, called as the radius of the inner caustic. It is given by:

$$r_s = \rho \sin \alpha, \tag{4.35}$$

For meridional rays, α and r_s are equal to zero. In Fig. 4.11, the collimated beam is focused obliquely at point Q on the fiber end face and makes an angle θ_s with the axis of the fiber. The point Q is at a distance of r_s from the fiber axis. The distance r_s, therefore, can also be written as:

$$r_s = f \tan \theta_s, \tag{4.36}$$

For $r_s = \rho$, angle θ_s will have maximum value and in that case maximum number of skew rays will be excited. In this case, Eq. (4.34) for the number of reflections will be modified to the following equation[7]:

$$N_{\text{ref}}(\theta, \alpha) = \frac{L}{2\rho \cos \alpha \tan \theta}, \tag{4.37}$$

While the angular power distribution inside the fiber (Eq. (4.31)) will become:

$$dP \propto \frac{n_1^2 \sin \theta \cos \theta}{\left(1 - n_1^2 \cos^2 \theta\right)^2} \cos \theta_s d\theta, \tag{4.38}$$

The transmitted power through the fiber in the presence of skew rays is then given by the following relation:

$$P_{\text{trans}} = \frac{\int_0^{\alpha_{\max}} \int_{\theta_{\text{cr}}}^{\pi/2} R_p^{N_{\text{ref}}(\theta,\alpha)} \frac{n_1^2 \sin\theta \cos\theta}{\left(1-n_1^2 \cos^2\theta\right)^2} \cos\theta_s d\theta d\alpha}{\int_0^{\alpha_{\max}} \int_{\theta_{\text{cr}}}^{\pi/2} \frac{n_1^2 \sin\theta \cos\theta}{\left(1-n_1^2 \cos^2\theta\right)^2} \cos\theta_s d\theta d\alpha}. \tag{4.39}$$

The above expression of power transmission includes the sensing length L, fiber core radius ρ and the numerical aperture of the fiber. Therefore, the change in the value of any of these parameters can affect the performance of the sensor. Figure 4.13 shows the variation of the sensitivity of the sensor with skewness parameter for different

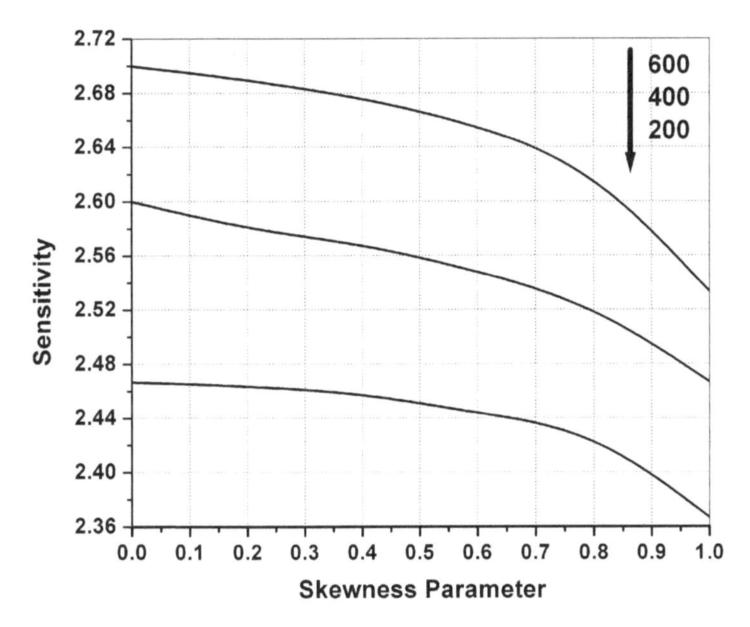

Fig. 4.13. Variation of sensitivity of the sensor with skewness parameter ($\sin \alpha$) for different values of fiber core diameter (2ρ, in micron). The results are for metallic film of silver over the core of the fiber. Reprinted from ref. [7] with permission from Optical Society of America.

values of fiber core diameter.[7] The *sine* of the skewness angle is termed as skewness parameter. It is observed that for a fiber of given core diameter, the sensitivity decreases with an increase in the skewness parameter. In other words, as the number of skew rays increases, the sensitivity decreases. The value of skewness parameter for meridional rays is equal to zero and hence the results imply that the sensitivity is the highest for meridional rays. Further, as the core diameter decreases, for a given skewness parameter, the sensitivity decreases. This is because as the core diameter decreases, the number of reflections in the sensing region increases, which decreases the sensitivity of the sensor. The sensitivity of the sensor also depends on the length of the sensing region. The variation of sensitivity with skewness parameter for different values of the sensing length (L) is shown in Fig. 4.14.[7] The figure shows that the sensitivity decreases with the increase in the sensing length of the sensing region. This is

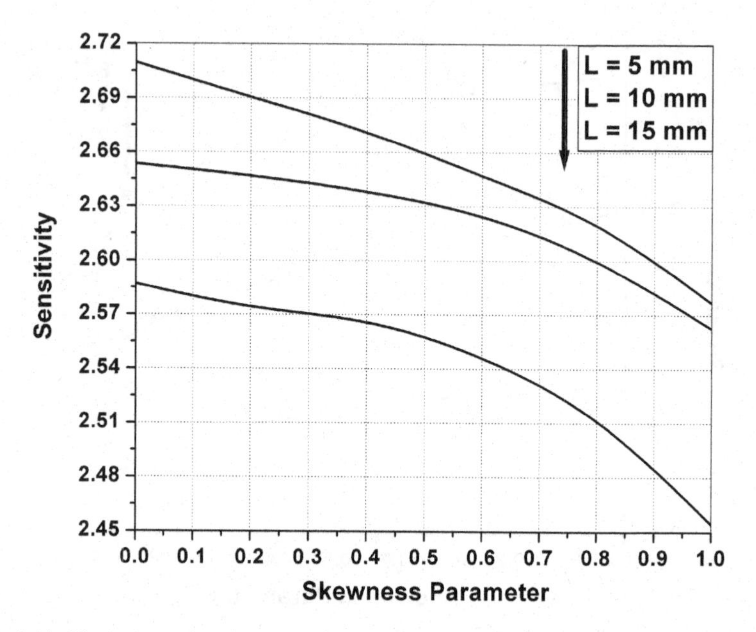

Fig. 4.14. Variation of sensitivity of the sensor with skewness parameter ($\sin \alpha$) for different values of the sensing length (L). The results are for metallic film of silver over core of the fiber. Reprinted from ref. [7] with permission from Optical Society of America.

because of the increase in the number of reflections as the length of the sensing region increases.

4.5 Diffuse Source

We now discuss the launching of light in the fiber using a diffuse source such as LED. This is quite relevant for miniaturization of sensors, as the bulky collimation and focusing optics with broadband source can simply be replaced by an LED placed at the end face of the optical fiber. The power distribution inside the fiber for the diffuse source is given by[6,8]:

$$dP \propto n_1^2 \sin\theta \cos\theta d\theta, \tag{4.40}$$

where, again n_1 is the refractive index of the fiber core and θ is the angle of the incident ray with the normal to the core-metal interface. For SPR-based fiber optic sensor utilizing diffuse source, there will be

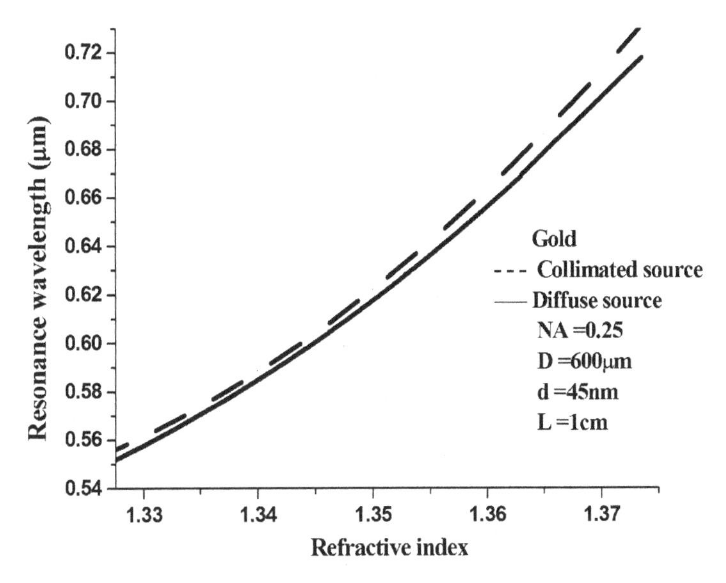

Fig. 4.15. Variation of resonance wavelength with the refractive index of the sensing medium for two different kinds of light sources.

both meridional and skew rays inside the fiber core. In the presence of both, the transmitted power through the fiber is given by[9]:

$$P_{\text{trans}} = \frac{\int_0^{\pi/2} d\alpha \int_{\theta_{\text{cr}}}^{\pi/2} R_p^{N_{\text{ref}}(\theta,\alpha)} n_1^2 \sin\theta \cos\theta d\theta}{\int_0^{\pi/2} d\alpha \int_{\theta_{\text{cr}}}^{\pi/2} n_1^2 \sin\theta \cos\theta d\theta}, \tag{4.41}$$

where $N_{\text{ref}}(\theta, \alpha)$ is given by Eq. (4.37). Equation (4.41) will convert to the transmitted power equation for meridional rays only if the skewness angle α is equal to zero. Figure 4.15 shows the variation of resonance wavelength with the refractive index of the sensing medium for the following two kinds of light launching: (i) collimated source and focusing lens combination assuming only meridional rays are propagating in the fiber and (ii) the diffuse source where meridional and skew rays both propagate in the fiber. The plots are for the gold metal layer of thickness 45 nm. The following values of the fiber and probe parameters are used for the simulation: NA of the fiber $= 0.25$, $L = 1$ cm and $\rho = 300\,\mu$m. In both the cases, as the refractive index increases the resonance wavelength increases. The variation is nonlinear and the difference between the results obtained for two kinds of source is small.

4.6 Performance Parameters: Sensitivity, Detection Accuracy, and Figure of Merit (FOM)

The performance of a fiber optic SPR sensor is analyzed in terms of the following parameters: sensitivity, detection accuracy, and FOM. For the best performance of the sensor, all these parameters should be as high as possible. The sensitivity of the sensor is defined as the shift in the resonance wavelength (in SPR curve) per unit change in the refractive index of the sensing medium around the metal coated optical fiber sensing probe. Let λ_{res} be the resonance wavelength corresponding to n_s refractive index of the sensing medium and $\lambda_{\text{res}} + \delta\lambda_{\text{res}}$ is the resonance wavelength corresponding to the refractive index $n_s + \delta n_s$ of the sensing medium, then the sensitivity of the sensor at n_s refractive index of the sensing medium is, mathematically, defined as:

$$S = \frac{\delta\lambda_{\text{res}}}{\delta n_s}. \tag{4.42}$$

To obtain the sensitivity, the calibration curve of the sensor, (i.e., the resonance wavelength versus refractive index curve) is plotted and the slope of the curve is determined for different values of the refractive index of the sensing medium. The second parameter of importance is the detection accuracy which is proportional to the inverse of the full width at half minimum (FWHM) of the SPR curve. The broader the SPR curve is, the poor is the detection accuracy. Detection accuracy tells us that how accurately the system can measure the resonance wavelength. If $\Delta\lambda_{0.5}$ is the FWHM of the SPR curve at 50% contrast in the transmitted power, then the detection accuracy (DA) is written as:

$$DA \propto \frac{1}{\Delta\lambda_{0.5}}. \tag{4.43}$$

The third important parameter is the FOM which takes into account the sensitivity of the sensor and the spectral width of the SPR curve at 50% transmission in SPR curve. It is defined as the ratio of the sensitivity to the FWHM of SPR curve and is given as:

$$FOM = \frac{S}{\Delta\lambda_{0.5}}. \tag{4.44}$$

Higher is the value of FOM better is the sensor. The other important performance parameter of a SPR sensor is its operating range which is the range of the sensing parameter that can be measured or sensed by the sensor. While the operating range of intensity measurement-based SPR sensors is naturally limited due to the limited width of the SPR dip, the operating range of angular and the wavelength interrogation-based SPR sensors may be made much wider. In principle, the operating range of these sensors is determined by the detection system, more specifically by the angular or spectral range covered by the optical system-angular position detector array or spectrum analyzer, respectively. In other words, it is the range of the sensing layer refractive index (n_s) that a sensor can detect for a given wavelength range of the detection system. For example, if the spectrum analyzer/spectrometer operates in the visible region only then the resonance wavelength can have the same range in the visible region and the corresponding refractive index range will be the operating range of the sensor. Thus, one should design a SPR sensor which can take full advantage of the operating range of the detection system.

We have discussed, earlier, two kinds of sources for launching light in the optical fiber. One, collimated source with focusing lens to launch guided meridional rays in the fiber and two, the diffuse source which launches both meridional and skew rays in the fiber. To show the effect of sources on the sensitivity, width of SPR spectrum and FOM of the SPR-based fiber optic sensor, the simulated results are plotted in Figs. 4.16 to 4.18. The simulations have been carried out on gold coated sensor with following parameters: gold film thickness = 45 nm, NA of the fiber = 0.25, L = 1 cm and ρ = 300 μm. The results show that as the refractive index increases sensitivity and

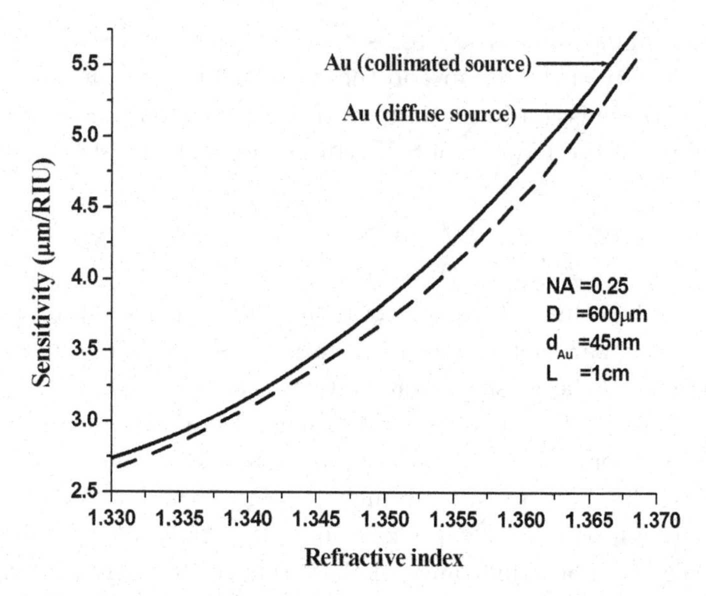

Fig. 4.16. Variation of sensitivity with the refractive index of the sensing medium for: (a) combination of collimated source and focusing lens, and (b) diffuse source.

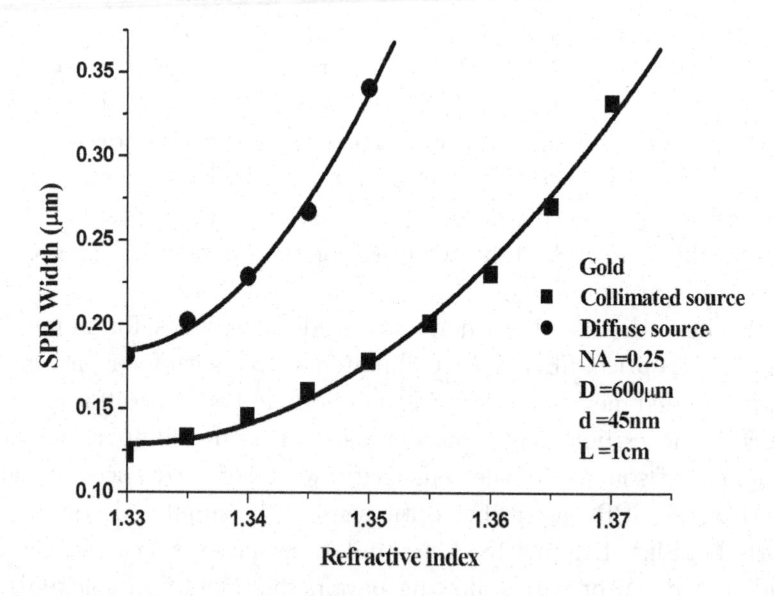

Fig. 4.17. Variation of width of SPR spectrum with the refractive index of the sensing medium for: (a) combination of collimated source and focusing lens, and (b) diffuse source.

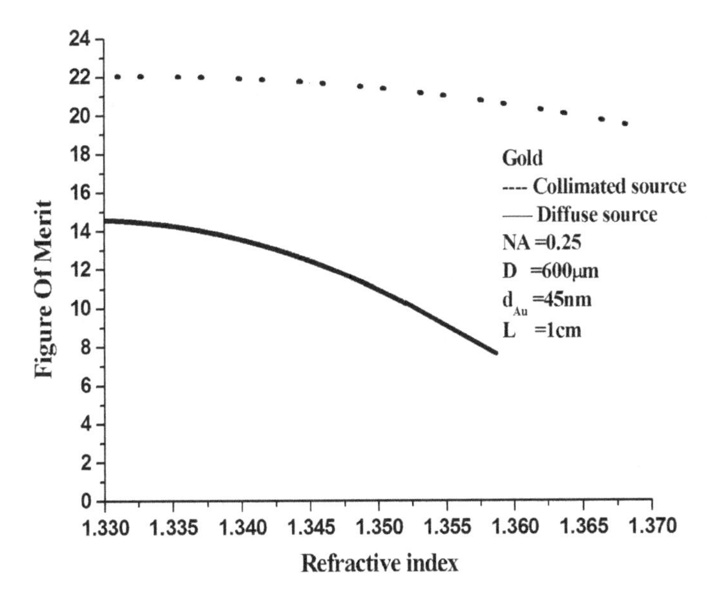

Fig. 4.18. Variation of FOM with the refractive index of the sensing medium for (a) combination of collimated source and focusing lens, and (b) diffuse source.

the width of the SPR spectrum increase, while the FOM marginally decreases irrespective of the source used for launching light. For a given refractive index, the sensitivity and the detection accuracy of the fiber optic SPR sensor using a combination of collimated source and focusing lens for light launching are better than the sensor that uses diffuse source for light launching. Thus the performance of the sensor will be better if maximum number of meridional rays is launched in the fiber.

4.7 Summary

In this chapter we have provided the method of modeling and simulation of a SPR-based fiber optic sensor for the detection of refractive index using N-layer model and matrix method. The modeling utilizes a two dimensional approach and geometrical optics. The parameters used to evaluate the performance of the sensor have also been discussed. The role of various kinds of light sources and their launching conditions on the performance of the sensor has also been described. In addition, the effects of refractive index of the sensing medium on the sensitivity and detection accuracy of the sensor have been shown.

References

1. S.K. Srivastava and B.D. Gupta, Fiber optic plasmonic sensors: past, present and future, *The Open Optics Journal* **7** (2013) 58–83.
2. T. Kimura, *Basic concepts of the optical waveguide in optical fiber Transmission* K. Noda (Ed), North Holland, 1986.
3. S.A. Maier, *Plasmonics: Fundamentals and applications*, Springer, 2007.
4. E. Hecht, *Optics*, Addison-Wesley Publishing Company, 1990.
5. B.D. Gupta and A.K. Sharma, Sensitivity evaluation of a multi-layered surface plasmon resonance based fiber optic sensor: A theoretical study, *Sensors and Actuators B* **107** (2005) 40–46.
6. A.W. Synder and J.D. Love, *Optical Waveguide Theory*, Chapman Hall, 1991.
7. Y.S. Dwivedi, A.K. Sharma and B.D. Gupta, Influence of skew rays on the sensitivity and signal to noise ratio of fiber-optic surface plasmon resonance sensor: a theoretical study, *Applied Optics* **46** (2007) 4563–4569.
8. B.D. Gupta and C.D. Singh, Evanescent-absorption coefficient for diffuse source illumination: uniform and tapered fiber sensors, *Applied Optics* **33** (1994) 2737–2742.
9. S. Singh, R.K. Verma and B.D. Gupta, LED based fiber optic surface plasmon resonance sensor, *Optical and Quantum Electronics* **42** (2010) 15–28.

Chapter 5

Fabrication and Functionalization Methods

We have already talked about the optical fiber sensors and what are they meant for. There are basically two types of the fiber optic sensors: one is based on the principle that all interaction occurs with the light guided internally and the modulations are observed in the signal at the output end; while in other type of sensors, the optical fiber works as an external device used for the coupling of light to the sensor. In the first kind of sensors, in general, the sensing surface is generated over the fiber surface itself. In the second kind of sensors, the light from the optical fiber is coupled to the sensing surface on some other substrate and the modulated light is coupled back to the optical fiber. The sensing surfaces are in general prepared with biomolecular recognition elements (BREs) specific to the analyte of interest. The interaction of the analyte or measurand with the BRE leads to a change in the overall morphology, structure, chemical composition, or refractive index on or in the vicinity of the sensor surface, which is further translated to a modulation in the interacting optical beam. In this chapter we shall present the methods of fabrication of the fiber optic surface plasmon resonance (SPR) sensors and the methods of functionalization of the sensor surfaces for specificity and accuracy. We shall first discuss the steps required for the fabrication of a fiber optic SPR probe and then we shall discuss various kinds of functionalization methods.

5.1 Sensing Elements

A sensor comprises various elements, called sensing elements, which are integral parts of any sensor. These include sensor surface, receptor layer, blocking agents, etc. We shall discuss these one by one.

5.1.1 *Sensor surface*

While discussing the sensor surface, we shall limit ourselves to the fabrication of fiber optic SPR sensors of the first kind which were discussed earlier. In such sensors the light beam guided into the core of the optical fiber gets modulated due to sensing phenomena at the surface of the core. The second kind of sensors in which the sensor surface is prepared over some other substrate surface other than the optical fiber is less relevant to discuss in the context of the present topic. However, the fabrication method remains mostly the same for both kinds of sensors. The fabrication steps of the fiber optic SPR sensor include the preparation of the optical fiber probe, cleaning and coating of the plasmonic metal/material of interest.

5.1.1.1 *Preparation of the fiber probe*

To prepare the fiber optic SPR probe the outer protective jacket of the fiber is stripped from the middle portion with the help of a sharp razor blade. Afterwards, the cladding is removed from this middle portion. In the case of optical fibers with glass cladding, the cladding is stripped by dipping the unjacketed portion in hydrofluoric acid (HF). This process needs lots of optimization, care in handling and expertise as HF is highly corrosive. The process of stripping of the cladding becomes much easier if the cladding of the optical fiber is made of plastic or polymer. In most of the modern day sensing studies, plastic clad silica (PCS) optical fibers are being preferred over conventional glass clad optical fibers. The cladding of such an optical fiber may easily be removed by a sharp razor blade. After the stripping, the core surface is rigorously cleaned with acetone and ethanol. It shall be useful to mention at this point that care should be taken in cleaning; otherwise the cladding may sometimes get dissolved/ruined in one of these solvents.

5.1.1.2 *Coating of the metal layer*

After the cleaning, the unclad portion is coated with a few nanometer thick layer of the desired plasmonic metal/material. Most general coating techniques are thermal evaporation and sputtering. In sputtering, a target of the coating material is generally eroded by a beam of highly energetic particles and hence the coating chamber requires less amounts of vacuum for coating. In thermal evaporation, relatively higher order of magnitudes of vacuum can be managed. Both the methods have their own advantages and disadvantages; however, the thermally evaporated thin metal films possess less contamination than the sputtered films due to high vacuums. This is one of the most desired conditions for SPR. Hence, in general, thermal evaporation technique is preferred over sputtering for coating of plasmonic metals. However, sputtered films possess better adhesion due to higher speed of the evaporated plume. In thermal evaporation, the coating material is melted by electrical current in a vacuum chamber and the evaporated flux obtained due to melting gets coated on the substrate. We discuss here the process of coating by thermal evaporation in detail.

The schematic of a standard coating unit utilizing thermal evaporation is shown in Fig. 5.1.

High vacuum is one of the most necessary requirements to coat a very smooth and good quality film on any substrate. Higher the vacuum, smaller the number of air molecules present in the chamber; which leads to negligibly small scattering of the evaporated metal plume, providing a directional deposition of the plume directly on the substrate. As shown in the diagram, the coating unit consists of a vacuum chamber connected to two vacuum pumps (rotary and diffusion). The coating is performed in this chamber only. The vacuum chamber has two valves, one is called gas inlet valve and the other is gas admittance valve. The first one is used for high tension (HT) cleaning while the other one is used for letting air in the chamber. A detailed working of these valves will be discussed

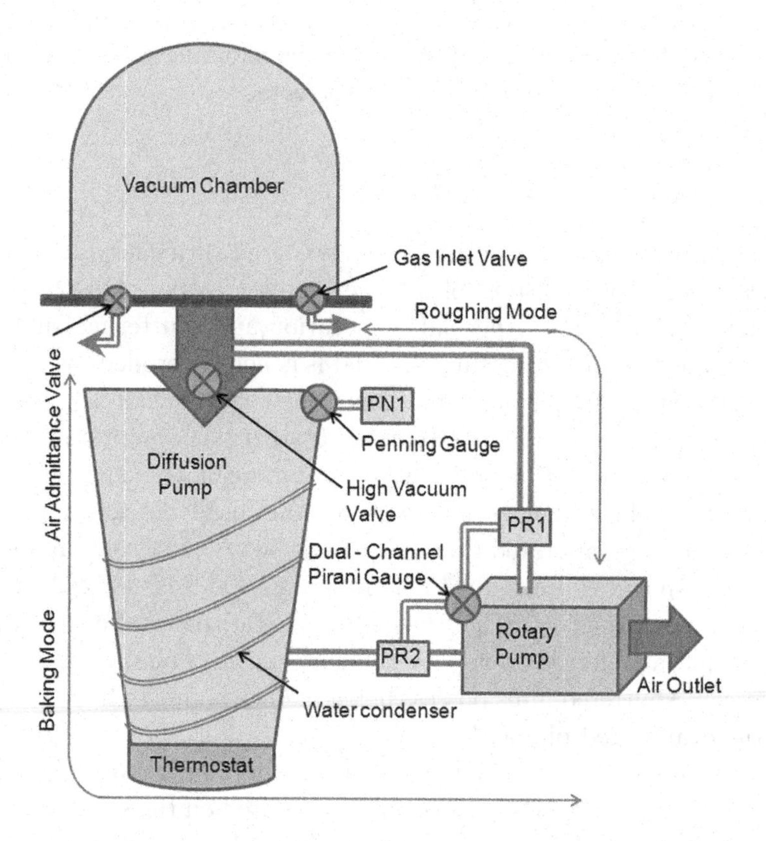

Fig. 5.1. A schematic of the coating unit.

in the following paragraphs at appropriate contexts. The vacuum chamber is connected to the rotary pump by two paths; one directly connected to it and working in the roughing mode, while the other via diffusion pump and working in baking mode. The vacuum chamber is evacuated primarily in roughing mode and then the coating unit is run in baking mode throughout the coating procedure. To achieve high vacuum, the diffusion pump is per-run in baking mode for about half an hour before the coating. The vacuum created by the rotary pump is measured by a dual channel Pirani gauge, which has two gauge heads; one (PR1) measuring vacuum in roughing mode and the other (PR2) in baking mode. The vacuum chamber is connected to the diffusion pump via a high vacuum valve, which is open only in baking mode. The vacuum chamber is primarily evacuated by the rotary pump in roughing mode and then by the diffusion pump when the high vacuum valve is open. The vacuum created by the diffusion pump is measured by a Penning gauge (PN1). The diffusion pump and the vacuum chamber are continuously cooled with a water condensing arrangement. A thermostat is also connected to the diffusion pump to monitor its temperature. If it gets too hot due to water flow failure or some other reasons, the thermostat switches the coating unit off. In modern days, many researchers prefer turbo pumps in place of diffusion pumps because of their oil free operation and relatively better vacuum by one order of magnitude. This also discards any possibility of oil contaminations in the coating. It however increases the cost manifold and the diffusion pumps serve the purpose to a great extent.

At first, the rotary pump is switched on in baking mode. The pressure of the pump is measured by the Pirani gauge head 2 (PR2). As soon as the pressure of the pump reaches below 5×10^{-2} mbar, the diffusion pump is switched on. The operation of rotary pump prior to the diffusion pump provides the necessary amount of vacuum to run the diffusion pump, as a vacuum of $\sim 10^{-3}$ mbar is required for the diffusion pump to start functioning. The diffusion pump along with the rotary pump is run for about half an hour. During this period, the pressure of the diffusion pump keeps on decreasing through the rotary pump.

Water is continuously run in the condenser throughout the coating to cool the diffusion pump as well as the vacuum chamber. The vacuum chamber is rigorously cleaned prior to the loading of the substrate and the coating material. Cleaner the vacuum chamber is, the easier it is to achieve a good vacuum. When the substrate (unclad optical fiber in the case of SPR sensor) is fixed inside the vacuum chamber and the coating material (generally gold/silver) is loaded on a tungsten filament/boat, the air admittance valve is closed and the coating unit is switched to roughing mode. In this mode of operation, the vacuum chamber is evacuated directly by the rotary pump. The pressure of the vacuum chamber in roughing mode is read by the Pirani gauge head 1 (PR1). As soon as the vacuum of the chamber reaches below 5×10^{-2} mbar, the machine is again switched to baking mode. The high vacuum valve is then slowly opened to let the chamber get connected to the diffusion pump. The pressure at this stage is measured by PN1. Since the pressure of the pump already reached a very low value due to half an hour pre-running before the coating, as soon as the chamber gets connected to the diffusion pump, its pressure decreases very rapidly. The pressure of the vacuum chamber reaches to 5×10^{-6} mbar in about 10 minutes. This is a very good vacuum for achieving a very smooth and good quality film of the coating material.

The inside view of the vacuum chamber is shown in Fig. 5.2. It consists of a base plate on which the boat/filament assembly and a substrate holder are mounted. A manual shutter is placed in between to control the coating by blocking the plume from boat/filament to the substrate. The boat/filament assembly is driven by a low tension (LT) current. Other components mounted on the base plate are a digital thickness monitor (DTM) and HT ion bombardment

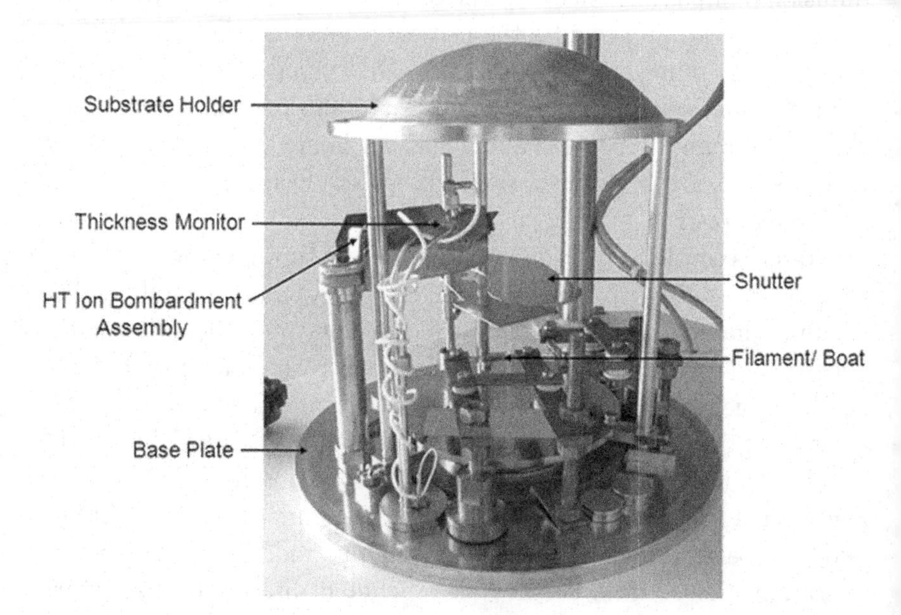

Fig. 5.2. Inside view of a typical vacuum chamber of the coating unit available in our laboratory (HINDHIVAC, model: 12A4D).

assembly. DTM is used for the online monitoring of the deposition rate and thickness of the film. It works on the principle of change in the oscillation frequency of the quartz crystal by deposited mass on it. The HT ion bombardment assembly is used for cleaning of the substrate before coating. For the cleaning purpose, the high vacuum valve is partially closed and the vacuum is slowly broken by opening the gas inlet valve and letting a stream of air in to attain a pressure of about $2 - 5 \times 10^{-2}$ mbar. Then the HT power supply is switched on and the current is increased slowly to reach a value of around 75 mA in the primary coil of HT transformer. Due to this current, a violet colored aurora is observed in the coating unit. This is in fact the plasma of air created by the HT current. This helps in the cleaning of the substrate, as the charged ions in the aurora remove any dust particles on the substrate. This process is carried out for 5 minutes. After that, the gas inlet valve is closed and the HT current supply is switched off. The high vacuum valve is again opened so that the pressure of chamber immediately reaches approximately to 5×10^{-6} mbar. The DTM unit is then switched on and density (DNT) and acoustic impedance (ACI) parameters for the coating material are fed into. The DTM unit consists of a quartz crystal microbalance which works on the inverse of piezoelectricity. It determines the mass per unit area by measuring the change in oscillation frequency of the quartz crystal. The quartz crystal consists of electrodes evaporated on both the sides. As the plume gets deposited on the crystal, a change in frequency occurs, leading to a change in the voltage, which is translated to the change in thickness and rate of deposition by electronic means taking consideration of the DNT and ACI parameters. Now the LT power supply is switched on to start the coating. The LT current is slowly increased to heat the filament/boat of the coating unit. With a rise in the LT current, the temperature and hence glow of the filament/boat increases and at a certain current/temperature the coating material starts melting. The LT current is fixed around this value and the required rate of deposition is achieved by further minor adjustments in the magnitude of LT current. In practice, a small deposition rate is found to provide better control and good quality coatings. After achieving the required deposition rate, the shutter is opened to let the plume of the coating material deposit on the substrate. As soon as the required thickness is obtained, the shutter is closed and the LT current is slowly decreased down to zero. The high vacuum valve is then closed and the coating unit is run in baking mode for another 15 minutes before switching it off. This maintains the high vacuum of the diffusion pump and helps achieving it easily when operated for the next time. The coated optical fibers are then taken out by passing air through the air admittance valve. After taking the fibers out, the vacuum chamber is again evacuated in roughing mode by the rotary pump.

The fabricated sensing probes can sense a variety of chemical and biological samples/analytes/processes. However, specific detection of the analyte of interest requires the functionalization of the sensor surface with certain BREs or receptors specific to the sample/analyte of interest. BREs/receptors are the specific elements having the capability of recognizing, binding, and sensing the presence of the analyte in the presence of a number of other similar chemical species at liquid or solid surface. There are several types of receptors; i.e., enzymes, aptamers, microbes, antibody, lipids, tissue, and organelles, etc. They have rigorously been discussed in Sec. 3.3.2 of Chapter 3.

We now discuss the functionalization of the fabricated sensor with these receptors to facilitate it with specificity and selectivity. Various kinds of sensor surfaces require different protocols for functionalization. We present in the succeeding paragraphs the importance and qualification of a suitable surface for choosing a particular protocol.

5.1.1.3 *Criterion for support selection*

For the immobilization/functionalization of biomolecules on a particular support, the very first step is the selection of a suitable protocol. It is a very crucial step because the immobilization process is different for different applications. Sometimes, it may happen other ways, when one has a freedom of support selection rather protocol.

Support selection also depends upon the biocatalyst of interest, molecule, environmental condition, etc. These are tough to choose but definitely there are few basic features which should be considered before the selection of support and suitable protocol.

- Support should be homogeneous and have chemical, physical, and biological stability.
- The protocol and the support both should be non-toxic and immune to environmental conditions, i.e., pH, temperature, and humidity.
- The support should be non-reactive to immobilized molecule and product of the final reaction.
- The support should have several functional groups to immobilize molecule and the protocol shall have good capability to hold them for a long time.
- Sometimes support selection demands the application specific characteristics for the support.

5.2 Immobilization Techniques

Immobilization is the fabrication of receptor layer on the sensor surface, which is an important part of any sensor. It is a technique which restricts the movements of the receptor molecule either completely or partially to an attachment of a solid surface. Number of biomolecules can be immobilized over support for fabricating the sensor. Mostly biomolecule immobilization is highly in use because of their applications in biological and biomedical fields. The immobilization of enzymes, antibody, antigen, peptides, drugs, nucleic acid, lipids, and others are possible nowadays. These all fall in the category of BREs. Thus, the immobilization of BRE and immobilization techniques plays key roles in the performance of the biosensors. The working capacity, resolution, stability, repeatability, response time, and other characteristics depend on the immobilization of the sensing elements. We will discuss in the following paragraphs various methods of immobilization of some known receptors of bio-sensing importance.

5.2.1 *Covalent binding*

In this immobilization process, covalent bonding occurs between the support and the biomolecules. It falls in the irreversible immobilization category. Since the covalent bonds are strong, this bonding holds the biomolecules strongly on the surface and hence leads to very low possibility of leakage or detachment of the immobilized enzyme/biomolecule. The absence of any barrier between the biomolecule and the support surface further leads to the stability of the immobilized biomolecule. There are two main ways of covalent immobilization; one is to attach the reactive functional group of the polymer/biomolecule in original form, while the second is the modification of the support surface for the generation of the active group on it, which further binds on the surface. In other words, a support matrix with active groups needs to be generated which can react with the functional groups of biomolecule to be attached on the surface. There are a number of methods for the immobilization through covalent binding, which depend upon the functional groups present in the enzyme, antibody, or other biomolecules and the nature of the support matrix. The amide bond formation, amidation reaction, Schiff base formation, N-ethyl-N-(3-dimethylaminopropyl carbodimide)(EDC)–N-hydroxysuccinimide (NHS) chemistry, thiol-disulfide interchange, and diazotization have repeatedly been used by many researchers. Covalent bonded surfaces are very much resistant to the changes in the temperature, ionic strength, pH etc. of the ambient. In most of the cases, they can be reused and recycled. Apart from these qualities, sensors utilizing covalent immobilization may suffer from potential loss of activity of the immobilized molecule. Therefore, it becomes quite essential to monitor the effect of coupling conditions, ambient temperature, pressure, humidity, and other important variants such as pH, etc. on the biological activity of the immobilizing molecule during the whole process. The schematic of the covalent immobilization is shown in Fig. 5.3. The molecules bind to the surface via covalent linkage.

Fig. 5.3. Covalent technique of immobilization.

5.2.1.1 *Thiol bonding*

Thiols are organo-sulphur compounds that contain carbon bonded sulphydryl group (-SH). They can also be called sulphur analogous of alcohols. Noble metals such as gold and silver have very high affinity towards strong thiol bonding. Thiols make coordinate covalent bonds (dative bonds) with metal ions to form metal thiolate complexes. The lone pair of electrons forming this dative bond comes from sulphur. The most common thiol compound used in SPR based sensors utilizing covalent bonding is 4-Aminothiophenol (4-ATP).

5.2.1.2 *Disulfide bonding*

This method is another form of the covalent bonding in which stable covalent bond is formed between modified support matrix and thiol groups. The method is also irreversible process as bonds can be easily burst by appropriate reagents. The activity of this method can be changed by changing the pH of the used thiol group which is its extra advantage. There are effective reagents, for example, maleimide and photonic induction. The method is quite stable. The disulfide binding on the support is shown in Fig. 5.4.

5.2.1.3 *Metal binding*

In this method, ligands immobilize over support matrix via bonding with metal ions. These ligands are weakly bonded with metal ions so that some specific amino acids can be used as alternative for this. This method is doing good as easy regeneration of the support

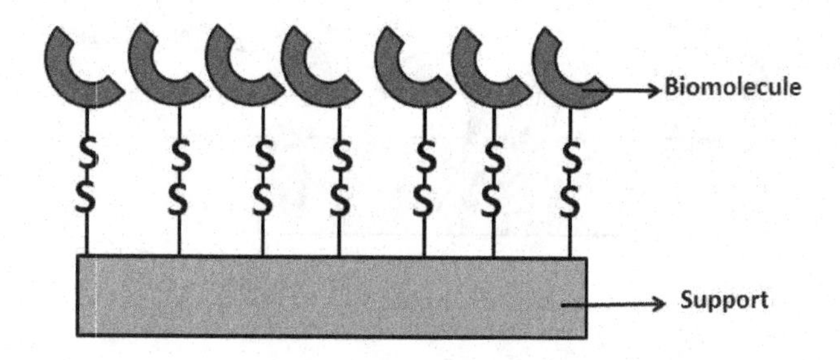

Fig. 5.4. Immobilization by disulphide bonding.

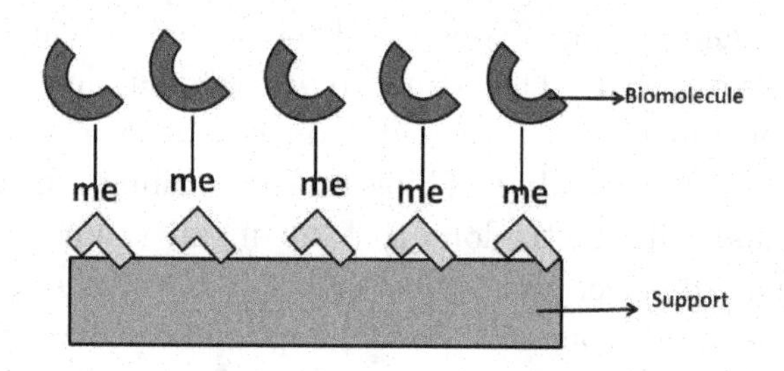

Fig. 5.5. Immobilization by metallic bonding.

matrix is possible without reducing the immobilization product. An illustration of metallic bonding is shown in Fig. 5.5.

5.2.1.4 *Silanization*

Silanization is the process of covalent bonding of various kinds of silanes over the surface of hydroxyls or oxides such as silica. In the case of silica, it is first treated with piranha solution (70:30, H_2SO_4: H_2O_2) to hydroxylate the surface. Various kinds of silanes such as aminosilanes, silanyl chloride, mercaptosilane are generally used depending upon the biomolecule/functional group of interest. For example, mercaptosilanes are used as adhesives between the glass surface and the metal; while the aminosilanes are used for the binding over glass surface to provide $-NH_2$ functional groups over the surface.

In general, the most used covalent linkages consist of thiol, hydroxyl, amines, and carboxylic acid functional groups. Biosensors generally utilize the covalent linkage of proteins such as enzymes and antibodies. Table 5.1 lists a few of the reported sensors utilizing covalent immobilization.

The covalent method of immobilization has many advantages, i.e., high stability, no leakage of molecules, and ease to immobilization, etc. In spite of these positive factors, however, it is rarely in use for cell immobilization due to the possibility of loss of enzymatic activity and degradation in performance with repeated use.

Table 5.1. A few examples of sensors utilizing covalent linkage.

S. No.	Immobilization protocol	Analyte	Reference
1	Glucose oxidase linked to aminosilanes immobilized on gold surface	Glucose	[1]
2	Anti-apolipoprotein B antibody modified by EDC–NHS chemistry and linked to 4-ATP attached to gold surface	Low density lipoprotein	[2]
3	Thiol derivatized oligonucleotide probes directly immobilized on gold surface	DNA	[3]
4	Cardiac troponin T antibody was immobilized on gold by a SAM of thiols by using cysteamine-coupling chemistry	Cardiac troponin T	[4]
5	Immobilization of azide-containing ligands by strain-promoted cycloaddition onto a cyclooctyne-modified SPR surface	Peptides and proteins	[5]
6	Anti-Palytoxin antibody attached covalently to SPR sensor surface	Palytoxin	[6]
7	Immobilization of a benzaldehyde — ovalbumin conjugate (BZ–OVA) on Au-thiolate SAMs containing carboxyl end groups	Benzaldehyde	[7]
8	Anti-pyraclostrobin monoclonal Antibodies attached to SPR surface	Pyraclostrobin	[8]
9	Anti-vitellogenin antibody attached to silver surface	Vitellogenin	[9]
10	T4 Bacteriophage covalently attached to silica surface via cross-linking	*E.coli*	[10]

(*Continued*)

Table 5.1. (*Continued*)

S. No.	Immobilization protocol	Analyte	Reference
11	Cocaine binding DNA aptamer	Cocaine in undiluted blood serum	[11]
12	Fluoresceinamine	pH	[12]
13	Glutamate Dehydrogenase (GDH)	Glutamate	[13]
14	Methylene blue labeled thrombin binding DNA aptamer (Oligore1)	Thrombin in blood plasma	[14]
15	Covalent coupling of respective conjugates to SAM of mercaptoundecanoic acid	DDT, chlorpyrifos, carbaryl	[15]
16	Immobilization of TNT analogs over OEG-thiol SAM	TNT	[16]
17	Covalent binding over Au-thiolate OEG monolayer	Insulin	[17]
18	Progesterone on CM–dextran layer	Progesterone	[18]
19	Anti-Hb antibody on CM–dextran layer	Hemoglobin	[19]
20	Myoglobin and Troponin specific antibodies on carboxymethylated dextran layer	Myoglobin, troponin	[20]
21	Protein conjugate on CM–dextran layer	Estradiol	[21]
22	Anti-folic acid antibody over dextran matrix	Folic acid (health supplement)	[22]
23	Antibody over Protein G layer	*Salmonella paratyphi*	[23]
24	Binding of digoxigenin-labelled ampicillin (DIG-AMPI) to PBP 2x* derivative of *Streptococcus pneumonia*	Penicillin and its derivatives	[24]

5.2.2 *Entrapment*

Entrapment is also an irreversible method where biomolecule is immobilized in the lattice structure or in polymer membrane. The lattice structure should be in right porosity/softness so that immobilized molecule would be stable and there is no leaching of

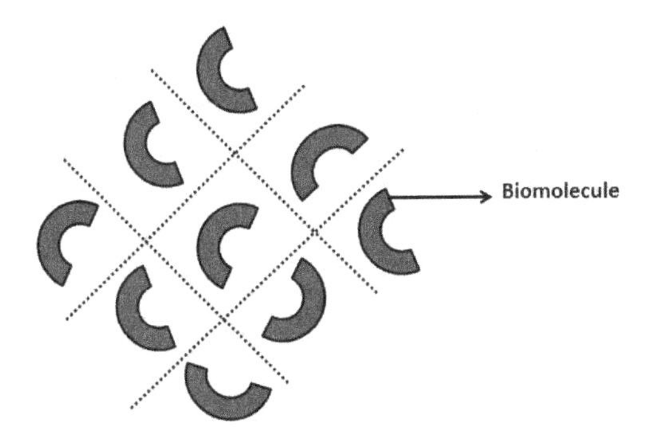

Fig. 5.6. Entrapment technique of immobilization.

its properties. This process provides good activity and stability to immobilized enzyme/molecule because there is no need of modification of the functional groups on the molecule. An illustration of immobilization of biomolecules by entrapment is shown in Fig. 5.6. The method works in mild environmental conditions (pH, polarity, temperature, etc.). The support matrix protects the immobilized molecules from contamination. Entrapment of molecules can be achieved with number of materials including: polymers, sol-gel, inorganic materials, alginate, silicon rubber, gelatine, polyvinyl alcohol, etc. The entrapment method gives high loading capacity but is costly and has diffusion limitations.

Few examples of studies utilizing the entrapment technique of immobilization are tabulated in Table 5.2.

5.2.3 *Encapsulation*

Encapsulation is also an irreversible process in which biomolecules are immobilized in semi permeable membrane. Molecular recognition element molecules freely float inside this semi-permeable membrane, while the analyte molecules can undergo bidirectional diffusion. Since all the encapsulations live within the membrane, there is less possibility of the contamination of biomolecules from the bad environmental conditions. The semi-permeable membrane also prevents the biocompatibility leakage, which results in stability of the biomolecule at support and hence the sensor response. The basic

Table 5.2. Examples of sensors based on entrapment technique of immobilization.

S. No.	Receptor	Analyte	Reference
1	Naringinase	Naringin	[25]
2	Acetylcholinesterase	Chlorphyrifos	[26]
3	Glucose Oxidase/urease	Glucose/urea	[27]
4	Tyrosinase	Phenolic compounds (phenol, catechol,m-cresol and 4-chlorophenol)	[28]
5	Lipase	Triglycerides	[29]
6	Alcohol dehydrogenase and nicotinic acid	Ethanol	[30]
7	Horseradish peroxidase	Hydrogen peroxide	[31]
8	Ruthenium complex [RII–tris (4,7-diphenyl-1,10-phenanthroline)]	Oxygen	[32]

requirement of this process is the proper pore size of the membrane and which should attenuate the outflow of the core material. Activity of the immobilized surface remains the same and lasts for long time because no modification is required in biomolecules during immobilization. The method is suitable for the immobilization of the microorganisms, living cells, enzymes, etc. Several biopolymers such as cellulose, nylon, alginate, chitosan, polyvinyl acetate, gelatine, boric acid, etc. are in use for encapsulations. A schematic of this method is shown in Fig. 5.7.

The encapsulation method is susceptible to contamination because the immobilization is performed within the membrane, and co-immobilization of several enzymes, cells with enzyme, etc. according to application is possible. The big disadvantage of this technique is that rigorous pore size control is needed and it is tough to manage for the case of small bio-molecular recognition elements, such as antibodies, etc. The most negative thing is that this is not suitable for the biomolecules which produce reaction products of similar size. We provide in Table 5.3 examples of some studies carried out on sensors utilizing encapsulation.

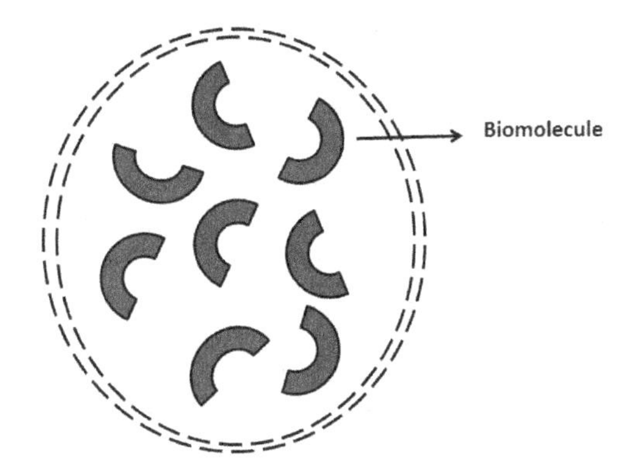

Fig. 5.7. Encapsulation technique of immobilization.

5.2.4 *Cross linking*

This method also falls in the category of the irreversible methods. It does not need any support mechanism for the immobilization, as shown in Fig. 5.8. Biomolecules are attached with one another by physical or chemical methods and form complex 3-dimensional matrix structures. Since, it does not need the support, so the process is cost-effective. Usually covalent bond forms between enzymes/biomolecules for the immobilization. For the immobilization, initially glutaraldehyde was used for enzyme aggregates; but it resulted in mechanical instability and low activity retention. Afterwards, it was improved by using crystalline enzymes with glutaraldehyde to overcome the problem of denaturation by heat and organic solvents and improve stability of the process along with reusability. But it was found that high purification of the enzymes is needed for cross-linking; so it is time consuming and laborious. An improvement was later suggested by using aqueous solution for the whole process which increases the stability and activity of the matrix while decreasing the labor and cost. Finally, the cross linking with enzyme aggregates in aqueous solution is found more applicable. This is advantageous over crystalline enzyme cross-links, but still it is constrained with the inhomogeneous size of obtained aggregates. There are several cross linking reagents that can be used other than glutaraldehyde according to the

Table 5.3. Examples of sensors utilizing encapsulation technique of immobilization.

S. No.	Recognition Element	Analyte	References
1	Tyrosinase	Phenol	[33]
2	Tris (1,10-phenanthroline) ruthenium chloride (Ru(phen)3) containing liposomes	Oxygen	[34]
3	Seminaphthorhodamine-1 carboxylate (SNARF-1C)	pH of blood	[35]
4	ruthenium complex, $[Ru(bpy)_3]\ Cl_2$	Nitrogen dioxide gas	[36]
5	Glucose oxidase	Glucose	[37]
6	Ruthenium bipyridyl complex encapsulated in zeolite Y	Oxygen	[38]
7	Concanavalin A and Dextran Encapsulated in a Poly (ethylene glycol) Hydrogel	Glucose	[39]
8	Silica nanoparticles encapsulated in graphene	Brest cancer biomarkers HER2 or EGFR	[40]

specific applications, i.e., dicorboxylic acid, bisisocyanate, diazonium salt, dextrane polysaccharide, and bovine serum albumin (BSA). The cross-linking method is support free, stable, and cheap but the utilization of the harsh chemicals makes it less applicable. Therefore, it is not very good for the immobilization, but it is mostly used for the enhancement of other immobilization methods. A number of studies utilizing cross-linking in sensor applications have been tabulated in Table 5.4.

5.2.5 *Adsorption*

Adsorption is a basic and very simple process of reversible immobilization based on the interaction between molecule and support matrix. Its use for the immobilization of cells and enzyme is very

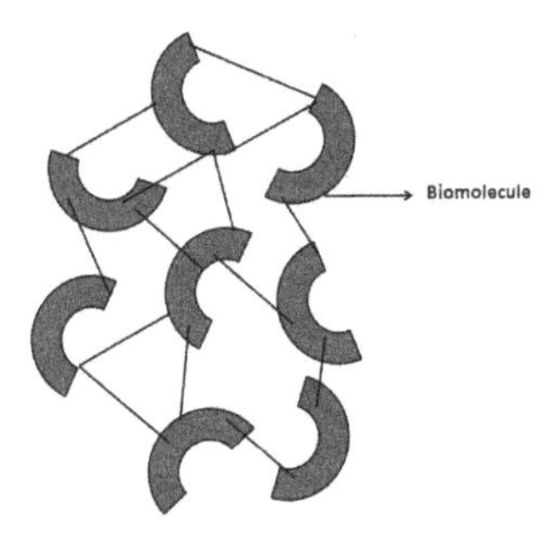

Fig. 5.8. Cross linking technique of immobilization.

Table 5.4. Examples of sensors utilizing cross-linking.

S. No.	Recognition Element	Analyte	References
1	Polyacrylic acid and polypyrene methanol (PAA–PM) with polyurethane	Metal ions (Fe^{2+}, Hg^{2+} and 2,4-dinitrotoluene	[41]
2	Anti-Staphylococcal enterotoxin B IgG (a-SEB)	Staphylococcal enterotoxin B IgG (a-SEB)	[42]
3	Polyphenol oxidase with chitosan and cyanuric chloride	Thiodicarb	[43]
4	Polyelectrolytes	Humidity	[44]
5	Glucose oxidase	Glucose	[45]
6	Cross-linked enzyme crystals (CLEC) of laccase	Phenol	[46]
7	Horseradish peroxidase	Hydrogen peroxide	[47]

often. The adsorption procedure involves two steps only; first is the attachment of molecules and support matrix under the suitable environmental conditions (pH, temperature, ionic strength) while mixing. Incubation is allowed for a particular period of time to let the molecules adsorb on the support matrix. Second step is the washing of the support to remove any unbound biomolecule from the surface once it dries after the first step. Adsorption is based on weak interactions. This process involves van der Waals forces, hydrogen bonding, ionic interaction, affinity binding, etc. The technique is simple, cost-effective, and fast, because it works in very modest conditions. It is reversible, which makes it possible for reuse. There is no chemical modification in support and biomolecule which makes it retentive. Apart from all these advantageous features, there are some drawbacks also, as it involves the weak interactions, the rate of leakage of biomolecules is high, loading capability of the support is random and the binding is unstable, which causes the problem in repeatability of the sensor response. Moreover, contamination of the substrate may also happen in adsorption due to fouling or non-specific bindings. This process is very susceptible to the environmental conditions (pH, temperature, humidity, ionic strength) which is the main reason for the leaching of biomolecules. A list of examples utilizing sensors based on adsorption is presented in Table 5.5.

Table 5.5. Examples of sensors based on adsorption.

S. No.	Recognition Element	Analyte	References
1	Glucose oxidase (GOD)	Glucose	[48]
2	Lactate oxidase	Lactate	[49]
3	Meldola's blue (MB)	Ethanol	[50]
4	Bromocresol Purple	Ammonia	[51]
5	Immobilization of 2,4-D-BSA conjugate	2,4-dichlorophenoxyacetic acid	[52]
6	HBP–mAb antibody-ELISA	2-hydroxybiphenyl (HBP)	[53]
7	Physisorption of TNPh–OVA conjugate	TNT	[54]
8	Au/BaP–BSA conjugate	Benzo[a]pyrene (BaP)	[55]
9	Immobilization of TNP–BSA conjugate	TNP	[56]
10	Physisorption of MA–albumin conjugate	Methamphetamine (MA)	[57]
11	Physisorption of Au/MO–BSA conjugate	Morphine	[58]
12	D_3 dopamine receptor (D-RC) and (DA–BSA) conjugate	Dopamine	[59]

5.3 Molecular Imprinting

As we have seen, there are varieties of methods for the immobilization of the biomolecules. Every biomolecule has its own specific qualities and capabilities to detect the signal. It is quite tough to find the most suitable/compatible biomolecule for the analyte/sample and then the most appropriate method for the preparation of the receptor. The molecular imprinting (MIP) is a technique that is very sophisticated and molecules are arranged in predetermined manner on the support matrix and each imprint has almost similar functions. All the molecules know each other and their functions. MIP provides highly selective and economic receptors. In this method, molecules are fixed in a polymeric membrane and form certain recognition cavity of similar structure after easy chemical processing. Therefore, these cavities are able to detect the same molecules which were settled earlier in polymeric membrane. MIP can be defined as a technique for the creation of binding sites with the memory of size, shape, and functional group of imprinted molecules (target molecules). Molecularly imprinted techniques have been shown to be an efficient way of providing functionalized materials and recognizing the specific molecules in a mixture of related compounds.

MIP is a very simple method. It requires few materials, i.e., functional monomer, target molecule, solvent, and cross linking agents. The three basic steps performed in MIP process are the (i) preparation of the polymerization solution, (ii) polymerization of the polymerization solution with target molecules, and (iii) removal of the target molecules. Polymerization is performed at certain temperature and monomers are arranged around the target molecules. Complete process results in the organization of the functional groups of the monomers in the desired places in the cavities through various interactions, i.e., electrostatic, hydrogen bonding, Vander Waal, polar, and non-covalent interactions. This makes its surface very specific to particular target molecule. Moreover, if there is any need to detect guest molecule then polymerization could be performed with the guest molecule with appropriate monomer and cross linker under appropriate conditions and ratios. There are two types of MIP in use for the creation of binding sites for target molecules. One is the covalent imprinting and the other is non-covalent imprinting.

5.3.1 *Covalent molecular imprinting*

In this type of imprinting, initially functional monomers carrying certain functional groups and targets are bound primarily to each other by covalent linkage before polymerization. Afterwards, this covalent conjugate is polymerized under certain conditions. After the polymerization, covalent linkage is broken and target molecule is removed from the polymer. Upon the exposure of this imprinted surface to target molecules, the same kind of covalent linkage is again formed, which makes changes in the surface properties. This method is modest to choose the ambient conditions (temperature, pH, etc.) because covalent linkage is very stable. A schematic of the preparation of the covalent MIP is shown in the Fig. 5.9.

5.3.2 *Non-covalent molecular imprinting*

In this type of imprinting, monomer and cross-linker with target molecules are synthesized together. This acts as molecular template. Monomers are arranged around the target molecules through non-covalent interaction (i.e., hydrogen binding, electrostatic interaction, and co-ordinate bond formation) as we have described earlier, this action is followed by a polymerization process with high degree of cross-linking. Polymerization process is followed by removal of these target molecules, which leaves certain recognition binding sites similar in shape, size and functional group of target molecules. These high receptor cavities can easily rebind with target molecules from

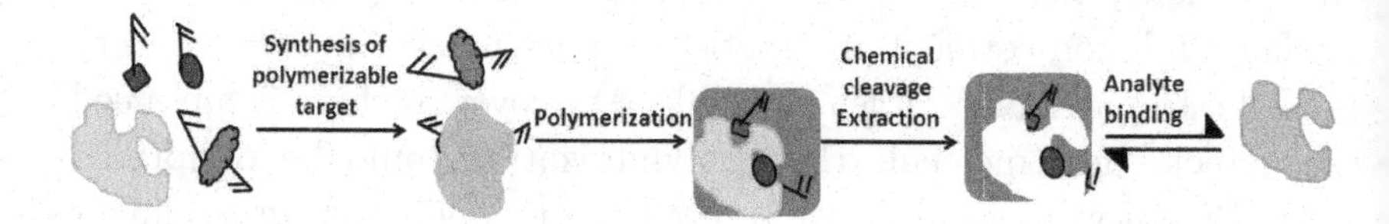

Fig. 5.9. Schematic of covalent imprinting.

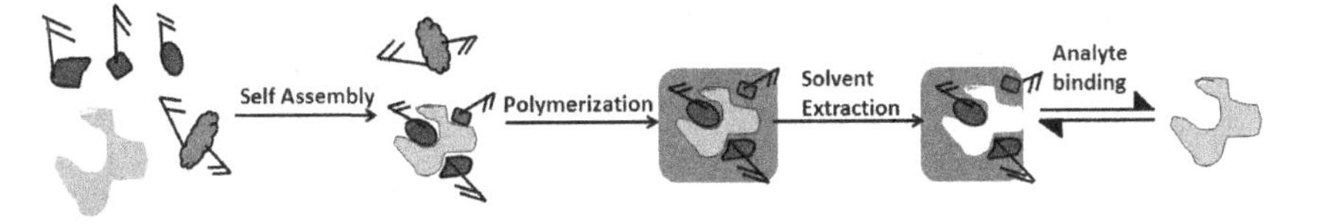

Fig. 5.10. Schematic of non-covalent imprinting.

closely related compounds mixture. The schematic of the preparation of the non-covalent imprinting is shown in Fig. 5.10.

Both the methods are used according to the requirements of the applications. One should choose either of these two methods, depending on the need and situation of their operations according to target molecules, selectivity, the cost, and time allowable for the preparation, etc. Both the methods have some basic requirements, advantages, and drawbacks as discussed in the succeeding paragraphs.

5.4 Advantages and Disadvantages of Molecular Imprinting

5.4.1 *Covalent imprinting*

Advantages

(1) A wide variety of polymerization conditions (e.g., high temperature, high or low pH, and highly polar solvent) can be employed, since the conjugates are formed by covalent linkages and are sufficiently stable.
(2) Monomer and target conjugate are stable, so the structure and binding sites are clear and easy to understand.

Disadvantages

(1) The method is troublesome and has costly synthesis of the monomer-template conjugate.
(2) Limited number of reversible covalent linkages.
(3) The binding of the target molecules is slow since they involve the formation of breakdown of covalent linkage.

(4) Sometimes breakdown of the covalent linkage happens during the binding and releasing of the guest molecules, which may result in false signal.

5.4.2 *Non-covalent imprinting*

Advantages

(1) The synthesis of monomer, cross linker, and target molecules is simple and easy as monomer-template conjugation is unnecessary.
(2) The templates can be removed easily from polymer under very mild conditions. The binding and releasing of target molecules is also easy and fast because of non-covalent interactions.

Disadvantages

(i) The polymerization conditions must be carefully chosen to maximize formation of non-covalent binding in the mixtures.
(ii) The process is not as clear as covalent process.
(iii) In this method functional monomers are used in large amount but sometimes it provides non-specific binding sites and target selectivity.

At last, the MIP surfaces are porous solids containing specific sites, which can interact with the molecule of interest according to a "key and lock" model. For this reason, a distinctive feature of MIPs, in comparison with other receptors, is the selectivity. The sensitivity of molecularly imprinted sensors can be determined by the amount of binding sites in the imprinted polymer, the electron transfer ability from the recognition sites to the electrode surface, and the electro-catalytic oxidation ability of the electrode to the template molecules. MIP has several other promising properties, i.e., high stability, reusability, low cost, and easy synthesis and offers considerable potential as a cost-effective alternative to the use of biomolecule/enzyme based recognition in various sensing applications. Imprinting can be used for small organic molecules such as pharmaceuticals, peptides, amino acids, pesticides, nucleotide bases, sugars, etc. Imprinting can also be used for the large organic molecules (proteins, cells) but it is a bit tough because of development of a large structure. Imprinting is also applicable for the metal and other ions sensing. Structure of MIP has to be optimized (ratios of the ingredients, functional monomer, target molecule, cross linker) according to different applications. Artificial molecularly imprinted receptors are also employed for several applications. Recently MIP with photonic crystals has been used for various studies. A number of examples utilizing MIP for sensing applications have been tabulated in Table 5.6.

5.5 Graphene Functionalized Receptors

Graphene is the thinnest material, discovered in 2004. It is a single atom thick sheet of sp_2 bonded carbon atoms arranged in a honeycomb lattice. Graphene has novel properties such that high electron mobility, high electrical conductivity, high young's modulus, large electrochemical potential window, lowest band gap, and lowest resistivity, etc. These promising properties open a window to the use of graphene in the sensing field, fabricating various devices, and batteries. Thin layer of graphene over a substrate makes it useful for several sensing applications.

To make any graphene based device, initially graphene or graphene oxide (GO) has to be synthesized. Secondly, one needs to check the compatibility of GO with the biomolecule. Then only it can be immobilized with several molecules/BREs/microbes/receptors according to the interest of the device and sensor. There are a number of molecules of biomedical interest but graphene and/or GO are not very compatible to biomolecules. For example, it was tested for microorganisms and found that cell adhesion power is decreased with the entrance of GO into cytoplasm.[87] Moreover, bacteria immobilized graphene sheets do not work, but after some cultural experiments/chemical processes it gives good results for the biomedical applications.[88] Graphene and GO should not be toxic for biomolecules for having good results for the sensing. Therefore, research is going on to make it more biocompatible and

Table 5.6. Examples of sensors based on MIP technique.

S. No.	Imprints	Analyte	Reference
1	Graphene oxide–SiO_2 composite MIP	Dopamine	[60]
2	PtNPs/GCE with MNA	17-β Estradiol	[61]
3	TEOS_Alumina gel MIPs	Catecholamines	[62]
4	Agarose immobilized MIP	Morphine	[63]
5	3-methacryloxypropyltrimethoxysilane modified silica particles MIP on GCE	Sulfonylurea Herbicide	[64]
6	Polystyrene photonic crystal template MIP	Amino acids	[65]
7	PMMA–polycarbazole hybrid poly [2-(N-carbazolyl) ethyl methacrylate-co-meth-acrylic acid MIP	l-Phenylalanine	[66]
8	SiO_2@Fe_3O_4 nanoMIPs with CNT	Tramadol in urine	[67]
9	Poly (2-hydroxyethylmethacrylate–methacryloylamidoaspartic acid) MIP	Kaempferol	[68]
10	Poly (2-hydroxyethylmethacrylate–methacryloylamidoglutamicacid) MIP	Tobramycin	[69]
11	MIPs from condensation of tetraethoxysilane (TEOS), phenyltriethoxysilane (PTEOS) and 3-aminopropyltrimethoxysilane (3-APTMS)	Caffeic Acid	[70]
12	Poly (vinyl chloride) (PVC)/nano-MIP	Urea	[71]
13	Phenol MIP on nitrogen doped graphene sheets	Methyl parathion	[72]
14	Acrylamide MIP on optical fiber	Cocaine	[73]
15	Methacrylic acid MIP on carbon paste electrode	Propylparaben	[74]
16	Poly (2-hydroxyethylmethacrylate–methacryloylamidoglutamic acid) MIP	Amoxicillin	[75]
17	Hydroxyl-terminated alkanethiol MIP	Carcinoembryonic antigen	[76]

(*Continued*)

Table 5.6. (*Continued*)

S. No.	Imprints	Analyte	Reference
18	Acrylamide MIP	Glyphosate	[77]
19	Terthiophene and carbazole MIP	Paracetamol, neproxen, theophyllin	[78]
20	3-aminophenylboronicacid (3-APBA) MIP	T-2 Toxin	[79]
21	Methacrylic acid (MAA) and trimethylolpropanetrimethacry late (TRIM) MIP	Organophosphorus pesticides	[80]
22	Polypyrrole MIP	2,4-dichlorophenoxy acetic acid	[81]
24	MIP from co-polymerization of the complex Cu(II)–catechol–urocanic acid ethyl ester with (tri)ethylene glycol dimethacrylate, and oligourethaneacrylate	Phenol	[82]
25	Acrylamide (AM) and trimethylol-propanetrimethacrylate (TRIM) MIP	L-Tryptophan	[83]
26	Methacrylamidohistidine (MAH) MIP	Glucose	[84]
27	Methacrylic acid (MAA) MIP	Hydroxyzine	[85]
28	Fluorescent phenylboronic acid functionalized mesoporous silica MIP	Saccharides	[86]

non-toxic for using this promising element at huge level in every field.

We shall give a brief overview for some graphene based sensing applications. First of all, GO is compatible with the nucleic acids and single stranded nucleic acids (DNA) gets adsorbed tightly on GO functionalized receptor. This beautiful feature is applied for the detection of hybridization processes.[89] A neural sensor using graphene has been recently reported, where fine graphene layers were first extracted by peeling of the graphite and then immobilized with the element of interest to make the sensor.[90] A cholesterol sensor has also been reported using graphene.[91] This biosensor was based on hybrid material receptor surface where receptor was first generated with graphene and Pt nanoparticles (PtNP) and then cholesterol oxidase and cholesterol esterase were immobilized over the receptor surface of graphene/PtNP. Graphene modified electrode was employed for the detection of glucose.[92] For this, the receptor surface was prepared by the combination of the nanocrystal of cadmium sulfide (CdS) with graphene layer which make highly effective sensor even for 0.7 mM concentration of glucose. Nicotinamide adenine dinucleotide (NADH) sensor was also reported with the help of graphene.[93] NADH, a co-enzyme, is a very important element for hundreds of dehydrogenase enzymatic reactions. This sensor was prepared by the non-covalent immobilization of graphene with methylene green over the receptor surface. Graphene has also been used for the detection of chromium ions (Cr) with gold nanoparticles.[94] Therefore, hundreds of applications of the graphene in the sensing and microelectronic devices have been employed.

The detailed immobilization of graphene and GO with other biocompatible elements over receptor surface is out of scope of this book. We limit ourselves only to a few examples of graphene based sensing applications. For more details, readers can consult following references.[95–98]

5.6 Summary

In this chapter we have presented various techniques that are used for the immobilization of biomolecules to fabricate the sensor. All the methods have their own advantages and disadvantages. Depending upon the requirement and the field of application of the sensor, a trade-off condition, a particular immobilization method is chosen. In most of the studies, covalent immobilization is preferred due to strong binding, which leads to stable and repeatable response for multiple uses. However, recently the interest in MIP based sensors has grown overwhelmingly due to its non-fouling properties. Further, a brief account of the sensors utilizing graphene has been presented.

References

1. S.K. Srivastava, R. Verma and B.D. Gupta, Surface plasmon resonance based fiber optic glucose biosensor, *Proceedings of SPIE* 8351 (2012) 83511Z.
2. R. Verma, S.K. Srivastava and B.D. Gupta, Surface-plasmon-resonance-based fiber-optic sensor for the detection of low-density lipoprotein, *Sensors Journal, IEEE* **12** (2012) 3460–3466.
3. R. Wang, S. Tombelli, M. Minunni, M.M. Spiriti and M. Mascini, Immobilisation of DNA probes for the development of SPR-based sensing, *Biosensors and Bioelectronics* **20** (2004) 967–974.
4. R.F. Dutra, R.K. Mendes, V. Lins da Silva and L.T. Kubota, Surface plasmon resonance immunosensor for human cardiac troponin T based on self-assembled monolayer, *Journal of Pharmaceutical and Biomedical Analysis* **43** (2007) 1744–1750.
5. A.E.M. Wammes, M.J.E. Fischer, N.J. de Mol, M.B. van Eldijk, F.P.J.T. Rutjes, J.C.M. van Hest and F.L. van Delft, Site-specific peptide and protein immobilization on surface plasmon resonance chips via strain-promoted cycloaddition, *Lab on a Chip* **13** (2013) 1863–1867.
6. B. Yakes, S. DeGrasse, M. Poli and J. Deeds, Antibody characterization and immunoassays for palytoxin using an spr biosensor, *Analytical and Bioanalytical Chemistry* **400** (2011) 2865–2869.
7. K.V. Gobi, K. Matsumoto, K. Toko, H. Ikezaki and N. Miura, Enhanced sensitivity of self-assembled-monolayer-based spr immunosensor for detection of benzaldehyde using a single-step multi-sandwich immunoassay, *Analytical and Bioanalytical Chemistry* **387** (2007) 2727–2735.
8. E. Mauriz, C. García-Fernández, J.V. Mercader, A. Abad-Fuentes, A.M. Escuela and L.M. Lechuga, Direct surface plasmon resonance immunosensing of pyraclostrobin residues in untreated fruit juices, *Analytical and Bioanalytical Chemistry* **404** (2012) 2877–2886.
9. S.K. Srivastava, A. Shalabney, I. Khalaila, C. Grüner, B. Rauschenbach and I. Abdulhalim, SERS biosensor using metallic nano-sculptured thin films for the detection of endocrine disrupting compound biomarker vitellogenin, *Small* **10** (2014) 3579–3587.
10. S.M. Tripathi, W.J. Bock, P. Mikulic, R. Chinnappan, A. Ng, M. Tolba and M. Zourob, Long period grating based biosensor for the detection of *escherichia coli* bacteria, *Biosensors and Bioelectronics* **35** (2012) 308–312.
11. J.S. Swensen, Y. Xiao, B.S. Ferguson, A.A. Lubin, R.Y. Lai, A.J. Heeger, K.W. Plaxco and H.T. Soh, Continuous, real-time monitoring of cocaine in undiluted blood serum via a microfluidic, electrochemical aptamer-based sensor, *Journal of the American Chemical Society* **131** (2009) 4262–4266.
12. C. Munkholm, D. Walt, F. Milanovich and S. Klainer, Polymer modification of fiber optic chemical sensors as a method of enhancing fluorescence signal for pH measurement, *Analytical Chemistry* **58** (1986) 1427–1430.

13. J. Cordek, X. Wang and W. Tan, Direct immobilization of glutamate dehydrogenase on optical fiber probes for ultrasensitive glutamate detection, *Analytical Chemistry* **71** (1999) 1529–1533.
14. Y. Xiao, A.A. Lubin, A.J. Heeger and K.W. Plaxco, Label-free electronic detection of thrombin in blood serum by using an aptamer-based sensor, *Angewandte Chemie International Edition* **44** (2005) 5456–5459.
15. E. Mauriz, A. Calle, A. Montoya and L.M. Lechuga, Determination of environmental organic pollutants with a portable optical immunosensor, *Talanta* **69** (2006) 359–364.
16. A. Larsson, J. Angbrant, J. Ekeroth, P. Månsson and B. Liedberg, A novel biochip technology for detection of explosives — TNT: synthesis, characterisation and application, *Sensors and Actuators B* **113** (2006) 730–748.
17. K.V. Gobi, H. Iwasaka and N. Miura, Self-assembled PEG monolayer based spr immunosensor for label-free detection of insulin, *Biosensors and Bioelectronics* **22** (2007) 1382–1389.
18. J.S. Mitchell, Y. Wu, C.J. Cook and L. Main, Sensitivity enhancement of surface plasmon resonance biosensing of small molecules, *Analytical Biochemistry* **343** (2005) 125–135.
19. S. Sonezaki, S. Yagi, E. Ogawa and A. Kondo, Analysis of the interaction between monoclonal antibodies and human hemoglobin (native and cross-linked) using a surface plasmon resonance (SPR) biosensor, *Journal of Immunological Methods* **238** (2000) 99–106.
20. J.F. Masson, L. Obando, S. Beaudoin and K. Booksh, Sensitive and real-time fiber-optic-based surface plasmon resonance sensors for myoglobin and cardiac troponin I, *Talanta* **62** (2004) 865–870.
21. M. Miyashita, T. Shimada, H. Miyagawa and M. Akamatsu, Surface plasmon resonance-based immunoassay for 17β-estradiol and its application to the measurement of estrogen receptor-binding activity, *Analytical and Bioanalytical Chemistry* **381** (2005) 667–673.
22. M. Boström Caselunghe and J. Lindeberg, Biosensor-based determination of folic acid in fortified food, *Food Chemistry* **70** (2000) 523–532.
23. B.-K. Oh, W. Lee, Y.-K. Kim, W.H. Lee and J.-W. Choi, Surface plasmon resonance immunosensor using self-assembled protein G for the detection of salmonella paratyphi, *Journal of Biotechnology* **111** (2004) 1–8.
24. G. Cacciatore, M. Petz, S. Rachid, R. Hakenbeck and A.A. Bergwerff, Development of an optical biosensor assay for detection of β-lactam antibiotics in milk using the penicillin-binding protein 2x*, *Analytica Chimica Acta* **520** (2004) 105–115.
25. Rajan, S. Chand and B.D. Gupta, Fabrication and characterization of a surface plasmon resonance based fiber-optic sensor for bittering component — naringin, *Sensors and Actuators B* **115** (2006) 344–348.
26. Rajan, S. Chand and B.D. Gupta, Surface plasmon resonance based fiber-optic sensor for the detection of pesticide, *Sensors and Actuators B* **123** (2007) 661–666.

27. R. Verma and B.D. Gupta, A novel approach for simultaneous sensing of urea and glucose by SPR based optical fiber multianalyte sensor, *Analyst* **139** (2014) 1449–1455.
28. S. Singh, S.K. Mishra and B.D. Gupta, SPR based fibre optic biosensor for phenolic compounds using immobilization of tyrosinase in polyacrylamide gel, *Sensors and Actuators B* **186** (2013) 388–395.
29. A. Baliyan, P. Bhatia, B.D. Gupta, E.K. Sharma, A. Kumari and R. Gupta, Surface plasmon resonance based fiber optic sensor for the detection of triacylglycerides using gel entrapment technique, *Sensors and Actuators B* **188** (2013) 917–922.
30. R. Verma and B.D. Gupta, Fiber optic surface plasmon resonance based ethanol sensor, *Proceedings of SPIE* 8992 (2014) 89920A.
31. Q. Deng and S. Dong, Mediatorless hydrogen peroxide electrode based on horseradish peroxidase entrapped in poly(o-phenylenediamine), *Journal of Electroanalytical Chemistry* **377** (1994) 191–195.
32. A.K. McEvoy, C.M. McDonagh and B.D. MacCraith, Dissolved oxygen sensor based on fluorescence quenching of oxygen-sensitive ruthenium complexes immobilized in sol-gel-derived porous silica coatings, *Analyst* **121** (1996) 785–788.
33. J. Yu, S. Liu and H. Ju, Mediator-free phenol sensor based on titania sol–gel encapsulation matrix for immobilization of tyrosinase by a vapor deposition method, *Biosensors and Bioelectronics* **19** (2003) 509–514.
34. K.P. McNamara and Z. Rosenzweig, Dye-encapsulating liposomes as fluorescence-based oxygen nanosensors, *Analytical Chemistry* **70** (1998) 4853–4859.
35. S.A. Grant and R.S. Glass, A sol–gel based fiber optic sensor for local blood ph measurements, *Sensors and Actuators B* **45** (1997) 35–42.
36. S.A. Grant, J.H. Satcher Jr. and K. Bettencourt, Development of sol–gel-based fiber optic nitrogen dioxide gas sensors, *Sensors and Actuators B* **69** (2000) 132–137.
37. P.C. Pandey, S. Upadhyay and H.C. Pathak, A new glucose sensor based on encapsulated glucose oxidase within organically modified sol–gel glass, *Sensors and Actuators B* **60** (1999) 83–89.
38. B. Meier, T. Werner, I. Klimant and O.S. Wolfbeis, Novel oxygen sensor material based on a ruthenium bipyridyl complex encapsulated in zeolite Y: dramatic differences in the efficiency of luminescence quenching by oxygen on going from surface-adsorbed to zeolite-encapsulated fluorophores, *Sensors and Actuators B* **29** (1995) 240–245.
39. R.J. Russell, M.V. Pishko, C.C. Gefrides, M.J. McShane and G.L. Coté, A fluorescence-based glucose biosensor using concanavalin a and dextran encapsulated in a poly(ethylene glycol) hydrogel, *Analytical Chemistry* **71** (1999) 3126–3132.
40. S. Myung, A. Solanki, C. Kim, J. Park, K.S. Kim and K.-B. Lee, Graphene-encapsulated nanoparticle-based biosensor for the selective detection of cancer biomarkers, *Advanced Materials* **23** (2011) 2221–2225.

41. X. Wang, C. Drew, S.-H. Lee, K.J. Senecal, J. Kumar and L.A. Samuelson, Electrospun nanofibrous membranes for highly sensitive optical sensors, *Nano Letters* **2** (2002) 1273–1275.
42. R. Slavık, J. Homola and E. Brynda, A miniature fiber optic surface plasmon resonance sensor for fast detection of staphylococcal enterotoxin B, *Biosensors and Bioelectronics* **17** (2002) 591–595.
43. F. de Lima, B.G. Lucca, A.M.J. Barbosa, V.S. Ferreira, S.K. Moccelini, A.C. Franzoi and I.C. Vieira, Biosensor based on pequi polyphenol oxidase immobilized on chitosan crosslinked with cyanuric chloride for thiodicarb determination, *Enzyme and Microbial Technology* **47** (2010) 153–158.
44. M.-S. Gong, J.-S. Park, M.-H. Lee and H.-W. Rhee, Humidity sensor using cross-linked polyelectrolyte prepared from mutually reactive copolymers containing phosphonium salt, *Sensors and Actuators B* **86** (2002) 160–167.
45. X. Yang, G. Johansson and L. Gorton, A glucose sensor made by chemically crosslinking glucose oxidase directly on the surface of a carbon electrode modified with Pd/Au for hydrogen peroxide electrocatalysis, *Microchimica Acta* **97** (1989) 9–16.
46. J.J. Roy, T.E. Abraham, K.S. Abhijith, P.V.S. Kumar and M.S. Thakur, Biosensor for the determination of phenols based on cross-linked enzyme crystals (clec) of laccase, *Biosensors and Bioelectronics* **21** (2005) 206–211.
47. H.-S. Wang, Q.-X. Pan and G.-X. Wang, A biosensor based on immobilization of horseradish peroxidase in chitosan matrix cross-linked with glyoxal for amperometric determination of hydrogen peroxide, *Sensors* **5** (2005) 266–276.
48. M. Ferreira, P.A. Fiorito, O.N. Oliveira Jr and S.I. Córdoba de Torresi, Enzyme-mediated amperometric biosensors prepared with the layer-by-layer (lbl) adsorption technique, *Biosensors and Bioelectronics* **19** (2004) 1611–1615.
49. V.G. Gavalas and N.A. Chaniotakis, Lactate biosensor based on the adsorption of polyelectrolyte stabilized lactate oxidase into porous conductive carbon, *Microchimica Acta* **136** (2001) 211–215.
50. A.S. Santos, R.S. Freire and L.T. Kubota, Highly stable amperometric biosensor for ethanol based on meldola's blue adsorbed on silica gel modified with niobium oxide, *Journal of Electroanalytical Chemistry* **547** (2003) 135–142.
51. P. Bhatia and B. Gupta, Surface plasmon resonance based fiber optic ammonia sensor utilizing bromocresol purple, *Plasmonics* **8** (2013) 779–784.
52. K.V. Gobi, S.J. Kim, H. Tanaka, Y. Shoyama and N. Miura, Novel surface plasmon resonance (spr) immunosensor based on monomolecular layer of physically-adsorbed ovalbumin conjugate for detection of 2,4-dichlorophenoxyacetic acid and atomic force microscopy study, *Sensors and Actuators B* **123** (2007) 583–593.
53. K.V. Gobi, H. Tanaka, Y. Shoyama and N. Miura, Continuous flow immunosensor for highly selective and real-time detection of sub-ppb levels of 2-hydroxybiphenyl by using surface plasmon resonance imaging, *Biosensors and Bioelectronics* **20** (2004) 350–357.

54. D. Ravi Shankaran, K. Matsumoto, K. Toko and N. Miura, Performance evaluation and comparison of four spr immunoassays for rapid and label-free detection of TNT, *Electrochemistry* **74** (2006) 141–144.
55. N. Miura, M. Sasaki, K.V. Gobi, C. Kataoka and Y. Shoyama, Highly sensitive and selective surface plasmon resonance sensor for detection of sub-ppb levels of benzo[a]pyrene by indirect competitive immunoreaction method, *Biosensors and Bioelectronics* **18** (2003) 953–959.
56. D.R. Shankaran, K.V. Gobi, K. Matsumoto, T. Imato, K. Toko and N. Miura, Highly sensitive surface plasmon resonance immunosensor for parts-per-trillion level detection of 2,4,6-trinitrophenol, *Sensors and Actuators B* **100** (2004) 450–454.
57. N. Miura, H. Higobashi, G. Sakai, A. Takeyasu, T. Uda and N. Yamazoe, Piezoelectric crystal immunosensor for sensitive detection of methamphetamine (stimulant drug) in human urine, *Sensors and Actuators B* **13** (1993) 188–191.
58. N. Miura, K. Ogata, G. Sakai, T. Uda and N. Yamazoe, Detection of morphine in ppb range by using SPR (surface-plasmon-resonance) immunosensor, *Chemistry Letters* **26** (1997) 713–714.
59. S. Kumbhat, D.R. Shankaran, S.J. Kim, K.V. Gobi, V. Joshi and N. Miura, A novel receptor-based surface-plasmon-resonance affinity biosensor for highly sensitive and selective detection of dopamine, *Chemistry Letters* **35** (2006) 678–679.
60. Y. Zeng, Y. Zhou, L. Kong, T. Zhou and G. Shi, A novel composite of S_iO_2-coated graphene oxide and molecularly imprinted polymers for electrochemical sensing dopamine, *Biosensors and Bioelectronics* **45** (2013) 25–33.
61. L. Yuan, J. Zhang, P. Zhou, J. Chen, R. Wang, T. Wen, Y. Li, X. Zhou and H. Jiang, Electrochemical sensor based on molecularly imprinted membranes at platinum nanoparticles-modified electrode for determination of 17β-estradiol, *Biosensors and Bioelectronics* **29** (2011) 29–33.
62. T.-R. Ling, Y.Z. Syu, Y.-C. Tasi, T.-C. Chou and C.-C. Liu, Size-selective recognition of catecholamines by molecular imprinting on silica–alumina gel, *Biosensors and Bioelectronics* **21** (2005) 901–907.
63. D. Kriz and K. Mosbach, Competitive amperometric morphine sensor based on an agarose immobilised molecularly imprinted polymer, *Analytica Chimica Acta* **300** (1995) 71–75.
64. P. Zhao and J. Hao, Tert-butylhydroquinone recognition of molecular imprinting electrochemical sensor based on core–shell nanoparticles, *Food Chemistry* **139** (2013) 1001–1007.
65. Y.-X. Zhang, P.-Y. Zhao and L.-P. Yu, Highly-sensitive and selective colorimetric sensor for amino acids chiral recognition based on molecularly imprinted photonic polymers, *Sensors and Actuators B* **181** (2013) 850–857.
66. Y. Chen, L. Chen, R. Bi, L. Xu and Y. Liu, A potentiometric chiral sensor for l-phenylalanine based on crosslinked polymethylacrylic acid–polycarbazole hybrid molecularly imprinted polymer, *Analytica Chimica Acta* **754** (2012) 83–90.

67. A. Afkhami, H. Ghaedi, T. Madrakian, M. Ahmadi and H. Mahmood-Kashani, Fabrication of a new electrochemical sensor based on a new nano-molecularly imprinted polymer for highly selective and sensitive determination of tramadol in human urine samples, *Biosensors and Bioelectronics* **44** (2013) 34–40.
68. V.K. Gupta, M.L. Yola and N. Atar, A novel molecular imprinted nanosensor based quartz crystal microbalance for determination of kaempferol, *Sensors and Actuators B* **194** (2014) 79–85.
69. M.L. Yola, L. Uzun, N. Özaltın and A. Denizli, Development of molecular imprinted nanosensor for determination of tobramycin in pharmaceuticals and foods, *Talanta* **120** (2014) 318–324.
70. A. Gültekin, G. Karanfil, M. Kuş, S. Sönmezoğlu and R. Say, Preparation of mip-based QCM nanosensor for detection of caffeic acid, *Talanta* **119** (2014) 533–537.
71. T. Alizadeh and A. Akbari, A capacitive biosensor for ultra-trace level urea determination based on nano-sized urea-imprinted polymer receptors coated on graphite electrode surface, *Biosensors and Bioelectronics* **43** (2013) 321–327.
72. X. Xue, Q. Wei, D. Wu, H. Li, Y. Zhang, R. Feng and B. Du, Determination of methyl parathion by a molecularly imprinted sensor based on nitrogen doped graphene sheets, *Electrochimica Acta* **116** (2014) 366–371.
73. S.P. Wren, T.H. Nguyen, P. Gascoine, R. Lacey, T. Sun and K.T.V. Grattan, Preparation of novel optical fibre-based cocaine sensors using a molecular imprinted polymer approach, *Sensors and Actuators B* **193** (2014) 35–41.
74. M.B. Gholivand, M. Shamsipur, S. Dehdashtian and H.R. Rajabi, Development of a selective and sensitive voltammetric sensor for propylparaben based on a nanosized molecularly imprinted polymer–carbon paste electrode, *Materials Science and Engineering: C* **36** (2014) 102–107.
75. M.L. Yola, T. Eren and N. Atar, Molecular imprinted nanosensor based on surface plasmon resonance: application to the sensitive determination of amoxicillin, *Sensors and Actuators B* **195** (2014) 28–35.
76. Y. Wang, Z. Zhang, V. Jain, J. Yi, S. Mueller, J. Sokolov, Z. Liu, K. Levon, B. Rigas and M.H. Rafailovich, Potentiometric sensors based on surface molecular imprinting: detection of cancer biomarkers and viruses, *Sensors and Actuators B* **146** (2010) 381–387.
77. P. Zhao, M. Yan, C. Zhang, R. Peng, D. Ma and J. Yu, Determination of glyphosate in foodstuff by one novel chemiluminescence–molecular imprinting sensor, *Spectrochimica Acta: Molecular and Biomolecular Spectroscopy* **78** (2011) 1482–1486.
78. R. Pernites, R. Ponnapati, M.J. Felipe and R. Advincula, Electropolymerization molecularly imprinted polymer (E-MIP) SPR sensing of drug molecules: pre-polymerization complexed terthiophene and carbazole electroactive monomers, *Biosensors and Bioelectronics* **26** (2011) 2766–2771.
79. G. Gupta, A.S.B. Bhaskar, B.K. Tripathi, P. Pandey, M. Boopathi, P.V.L. Rao, B. Singh and R. Vijayaraghavan, Supersensitive detection of t-2 toxin by the in situ synthesized π-conjugated molecularly imprinted nanopatterns.

An in situ investigation by surface plasmon resonance combined with electrochemistry, *Biosensors and Bioelectronics* **26** (2011) 2534–2540.
80. C. Wei, H. Zhou and J. Zhou, Ultrasensitively sensing acephate using molecular imprinting techniques on a surface plasmon resonance sensor, *Talanta* **83** (2011) 1422–1427.
81. C. Xie, S. Gao, Q. Guo and K. Xu, Electrochemical sensor for 2,4-dichlorophenoxy acetic acid using molecularly imprinted polypyrrole membrane as recognition element, *Microchimica Acta* **169** (2010) 145–152.
82. T.A. Sergeyeva, O.A. Slinchenko, L.A. Gorbach, V.F. Matyushov, O.O. Brovko, S.A. Piletsky, L.M. Sergeeva and G.V. Elska, Catalytic molecularly imprinted polymer membranes: development of the biomimetic sensor for phenols detection, *Analytica Chimica Acta* **659** (2010) 274–279.
83. F. Liu, X. Liu, S.-C. Ng and H.S.-O. Chan, Enantioselective molecular imprinting polymer coated QCM for the recognition of l-tryptophan, *Sensors and Actuators B* **113** (2006) 234–240.
84. A. Ersöz, A. Denizli, A. Özcan and R. Say, Molecularly imprinted ligand-exchange recognition assay of glucose by quartz crystal microbalance, *Biosensors and Bioelectronics* **20** (2005) 2197–2202.
85. M. Javanbakht, S.E. Fard, A. Mohammadi, M. Abdouss, M.R. Ganjali, P. Norouzi and L. Safaraliee, Molecularly imprinted polymer based potentiometric sensor for the determination of hydroxyzine in tablets and biological fluids, *Analytica Chimica Acta* **612** (2008) 65–74.
86. J. Tan, H.-F. Wang and X.-P. Yan, Discrimination of saccharides with a fluorescent molecular imprinting sensor array based on phenylboronic acid functionalized mesoporous silica, *Analytical Chemistry* **81** (2009) 5273–5280.
87. O. Akhavan and E. Ghaderi, Photocatalytic reduction of graphene oxide nanosheets on TiO_2 thin film for photoinactivation of bacteria in solar light irradiation, *The Journal of Physical Chemistry C* **113** (2009) 20214–20220.
88. H. Chen, M.B. Müller, K.J. Gilmore, G.G. Wallace and D. Li, Mechanically strong, electrically conductive, and biocompatible graphene paper, *Advanced Materials* **20** (2008) 3557–3561.
89. Y. Wang, Z. Li, D. Hu, C.-T. Lin, J. Li and Y. Lin, Aptamer/graphene oxide nanocomplex for in situ molecular probing in living cells, *Journal of the American Chemical Society* **132** (2010) 9274–9276.
90. C. Chiu, X. He and H. Liang, Surface modification of a neural sensor using graphene, *Electrochimica Acta* **94** (2013) 42–48.
91. R.S. Dey and C.R. Raj, Development of an amperometric cholesterol biosensor based on graphene–Pt nanoparticle hybrid material, *The Journal of Physical Chemistry C* **114** (2010) 21427–21433.
92. K. Wang, Q. Liu, Q.-M. Guan, J. Wu, H.-N. Li and J.-J. Yan, Enhanced direct electrochemistry of glucose oxidase and biosensing for glucose via synergy effect of graphene and CdS nanocrystals, *Biosensors and Bioelectronics* **26** (2011) 2252–2257.
93. H. Liu, J. Gao, M. Xue, N. Zhu, M. Zhang and T. Cao, Processing of graphene for electrochemical application: Noncovalently functionalize

graphene sheets with water-soluble electroactive methylene green, *Langmuir* **25** (2009) 12006–12010.
94. C. Santhosh, M. Saranya, R. Ramachandran, S. Felix, V. Velmurugan and A. Nirmala Grace, Graphene/gold nanocomposites-based thin films as an enhanced sensing platform for voltammetric detection of Cr(vi) ions, *Journal of Nanotechnology* 2014 (2014) 304526.
95. V. Georgakilas, M. Otyepka, A.B. Bourlinos, V. Chandra, N. Kim, K.C. Kemp, P. Hobza, R. Zboril and K.S. Kim, Functionalization of graphene: covalent and non-covalent approaches, derivatives and applications, *Chemical Reviews* **112** (2012) 6156–6214.
96. Y. Shao, J. Wang, H. Wu, J. Liu, I.A. Aksay and Y. Lin, Graphene based electrochemical sensors and biosensors: a review, *Electroanalysis* **22** (2010) 1027–1036.
97. Y. Zhu, S. Murali, W. Cai, X. Li, J.W. Suk, J.R. Potts and R.S. Ruoff, Graphene and graphene oxide: synthesis, properties, and applications, *Advanced Materials* **22** (2010) 3906–3924.
98. S.M. Kang, S. Park, D. Kim, S.Y. Park, R.S. Ruoff and H. Lee, Simultaneous reduction and surface functionalization of graphene oxide by mussel-inspired chemistry, *Advanced Functional Materials* **21** (2011) 108–112.

Chapter 6

SPR based Sensing Applications

6.1 Introduction

In the beginning of the book, the surface plasmon resonance (SPR) sensors were prism based where the evanescent wave exciting surface plasmons at the metal–dielectric interface resulted due to the total internal reflection of the incident beam at the prism–metal interface under the condition of angle of incidence greater than the critical angle. The prism based SPR sensors have few shortcomings like their bulky size, use of various optical and mechanical components, their inability of remote sensing, etc. Due to these shortcomings, replacement of prism by an optical fiber was suggested. This was because the propagation of light through optical fiber occurs because of the total internal reflection of the beam at the core–cladding interface. This means that the evanescent wave is also present in an optical fiber which propagates along the core–cladding interface. Therefore, the core of an optical fiber can replace the prism to design a SPR based fiber optic sensor. To fabricate a SPR based fiber optic sensor, plastic clad silica fiber is generally used because the removal of the cladding from the plastic clad silica fiber is very easy. Two kinds of fiber optic SPR probes have been designed, one in which the cladding is removed from a small portion of the fiber in the middle and in the other, the cladding is removed from one of the ends of the fiber and the end-face is made reflecting for the light to return back. The unclad core of the fiber is then coated with a metal layer. The metal layer is further surrounded by a dielectric sensing medium as shown in Fig. 4.1. In a SPR based fiber optic sensor, the propagating rays have a fixed range of angle of incidence and hence angle of incidence, like in a prism based SPR sensor, cannot be varied to find the resonance parameter. In other words, angular interrogation method is not useful for a SPR based fiber optic sensor and hence spectral interrogation method is used. For spectral/wavelength interrogation method one requires a polychromatic source for the excitation of surface plasmons at the metal–dielectric sensing medium interface. Thus, the light from a polychromatic source, using suitable optics, is coupled into one of the ends of the optical fiber SPR probe. The evanescent wave generated at the core–metal interface by the guided/propagating ray excites the surface plasmons at the metal–dielectric sensing medium interface. The coupling of evanescent wave with surface plasmons strongly depends on various factors such as wavelength, fiber parameters, probe geometry, metal layer properties, and sensing medium. In the case of prism based SPR sensor, there is only one reflection of the incident beam while in the case of a SPR based fiber optic sensor, the number of reflections for most of the guided rays is more than one. As the angle of incidence at the interface decreases, the number of reflections per unit length in the fiber increases. For a given ray, the number of reflections depends on the length of the probe/sensing region and also on the fiber core diameter. Increase in length or the decrease in the core diameter increases the number of reflections. Increasing the number of reflections increases the width of the SPR curve which is a disadvantage because an increase in the width of the SPR curve decreases the detection accuracy of the sensor. The spectrum of the light transmitted after passing through the SPR sensing region is recorded at the other end of the fiber. The SPR spectrum recorded for a given sensing medium around the metal film possesses shape similar to that shown in Fig. 4.8. The sensing is accomplished by determining the wavelength of the dip in the spectrum which has been named resonance wavelength in previous chapters. The resonance wavelength depends on the refractive index of the dielectric/sensing medium. The resonance wavelength is generally plotted as a function of the refractive index of the sensing medium, called the calibration curve of the SPR sensor, and the plot is used to find the sensitivity of the sensor. It is also used to know the refractive index of the unknown sensing medium. The sensitivity and the detection accuracy are determined in the same way as in the case of angular interrogation. The angles are replaced by wavelengths in the definitions of sensitivity and detection accuracy.

In Sec. 1.3 we have mentioned, in brief, the progress made in the field of SPR based fiber optic sensors. In this chapter, we shall describe some of the fiber optic sensors in more detail with their principle, application, and results.

6.2 Refractive Index Sensor

The fabrication of the SPR based fiber optic probe for refractive index sensing is the simplest. The experimental setup of the SPR based fiber optic refractive index sensor reported is shown in Fig. 6.1.[1] The setup consists of a flow cell to fix the fiber optic SPR probe and for filling the fluid for sensing, a polychromatic light source, a microscope objective for the coupling of light in the fiber probe, spectrometer to record the spectrum, and a personal computer interfaced with the spectrometer. Number of probes with 40 nm thick silver layer and varying thicknesses of silicon over layer were fabricated on a 1 cm unclad length of a 600 μm core diameter and 0.4 numerical aperture plastic clad silica fiber. One of these probes was fixed in the glass flow cell for measurements. Liquid samples of different refractive indices prepared using sucrose in de-ionized water were used for the calibration of the sensor.

Figure 6.2 shows the SPR spectra recorded for fiber optic SPR probe with silver/silicon films of 40 nm/5 nm thicknesses over an unclad fiber for different refractive indices of the fluid sample around the probe. The figure indicates the following: (i) each spectrum possesses a dip at a particular wavelength called as the resonance wavelength, and (ii) the SPR curve shifts towards the higher wavelength as the refractive index of the fluid increases. This is

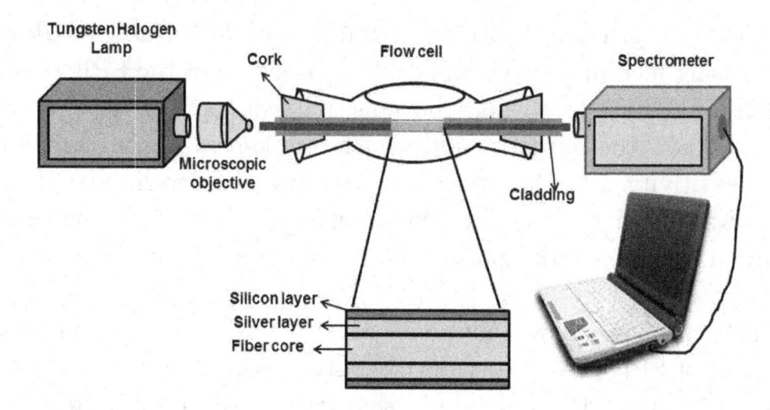

Fig. 6.1. Schematic of the experimental setup of a SPR based fiber optic refractive index sensor.

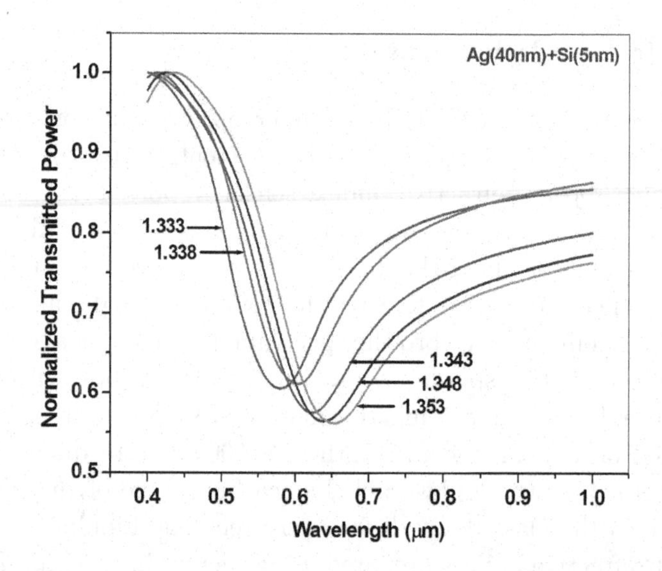

Fig. 6.2. SPR spectra of a fiber optic SPR probe with silicon layer of 5 nm thickness for refractive index of sensing region ranging from 1.333 to 1.353. Reprinted from Ref. [1] with permission from Optical Society of America.

in agreement with the first experimental study carried out on SPR based fiber optic sensor.[2] The only difference is the addition of a silicon layer over silver layer. The role of silicon layer was understood when these experiments were performed on probes with different thicknesses of the silicon layer over the same thickness of the silver

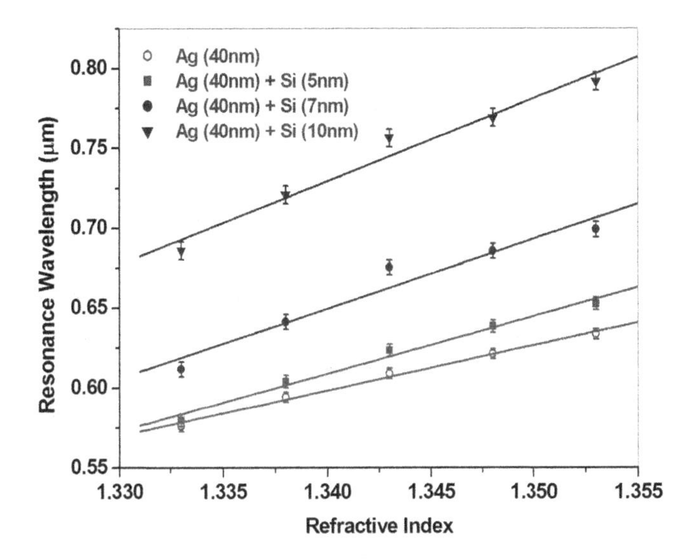

Fig. 6.3. Variation of resonance wavelength with the refractive index of the sensing layer for different thickness of silicon layer. Reprinted from Ref. [1] with permission from Optical Society of America.

layer. The resonance wavelengths determined from the SPR spectra were plotted as a function of refractive index of the sensing medium for each probe (Fig. 6.3). Following conclusions were drawn from these plots: (i) for a fixed refractive index of the sensing medium, the increase in the silicon layer thickness increases the resonance wavelength implying that the region of resonance wavelength can be tuned by adding the silicon layer, and (ii) the slope of the calibration curve (or the sensitivity) increases as the thickness of the silicon layer increases. However, the silicon layer thickness cannot be increased indefinitely because the increase in the thickness of silicon layer broadens the SPR curve and reduces the detection accuracy. The sensitivity enhancement occurs due to the enhancement of field at the silicon-sensing medium interface.[3]

6.2.1 *Effect of oxide layers*

The SPR based fiber optic sensor for the refractive index detection using additional layers other than silicon over metallic coating have

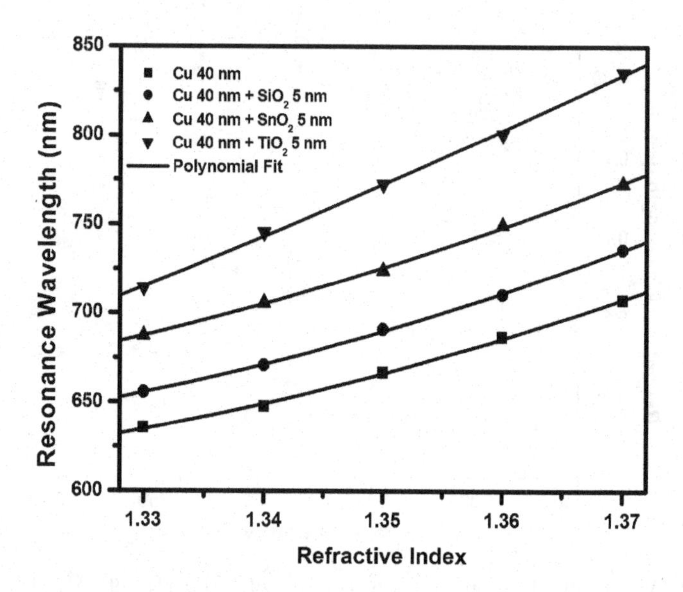

Fig. 6.4. Experimental variation of shift in the resonance wavelength for different oxide layer with refractive index of the sensing medium. Reprinted from Ref. [4] with permission from Elsevier.

also been reported.[4] The dielectric materials chosen for coating are silicon dioxide (SiO_2), tin dioxide (SnO_2), and titanium dioxide (TiO_2). These oxide materials are in the order of increasing dielectric constant. The experimental results obtained for 5 nm thick layer of SiO_2, SnO_2, and TiO_2 layer over 40 nm thick copper layer are shown in Fig. 6.4. In the same figure, the results for the fiber optic probe with 40 nm thick copper layer only are also shown. Following conclusions were drawn from the results plotted in Fig 6.4: (i) for a given refractive index, resonance wavelength depends on the material of the oxide layer. Increase in the dielectric constant of the oxide layer increases the resonance wavelength implying that the suitably chosen oxide layer can tune region of the resonance wavelength, and (ii) since the slope of each curve is different and increases from bottom to top it implies that the increase in the dielectric constant of the oxide layer increases the sensitivity of the sensor. Again, the sensitivity enhancement occurs due to the increase in the evanescent field at the interface of the oxide layer and the sensing medium.

6.2.2 *Multi-channel sensing*

The refractive index sensors discussed earlier are single channel sensor that means these can detect refractive index at one particular region or can detect refractive index of only one sample at a time. The SPR based fiber optic probe for detecting refractive index at multiple places or of multiple samples using same set up has been recently reported.[5] The schematic of three channels fiber optic SPR sensor is shown in Fig. 6.5. The probe consists of three different regions on a multimode optical fiber coated with either single or combination of different materials. In the probe shown in Fig. 6.5, channel 1, channel 2, and channel 3 have metallic coatings of silver, gold and copper, respectively while channel 3 has a thin over-layer of TiO_2 over copper layer. For performing experiments all the channels were separated by partitions. The samples of same or different refractive indices were kept around the channels for the calibration and testing of the probe.

Figure 6.6 shows the SPR spectra for a number of aqueous solutions around the three sensing regions. In Fig. 6.6(a) all the three channels have same sample while in Fig. 6.6(b) all the three channels have different samples. One thing which can be observed from these figures is that three dips exist in the SPR curves corresponding to three sensing channels and are well separated. These three dips are due to different coatings used for the three channels. Further, similar to one channel sensor, the dips shift towards higher wavelength with

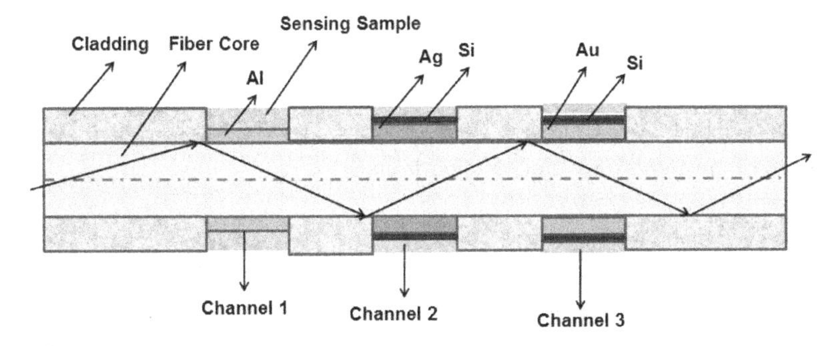

Fig. 6.5. Schematic of proposed three channels fiber optic SPR probe. Reprinted from Ref. [5] with permission from SPIE.

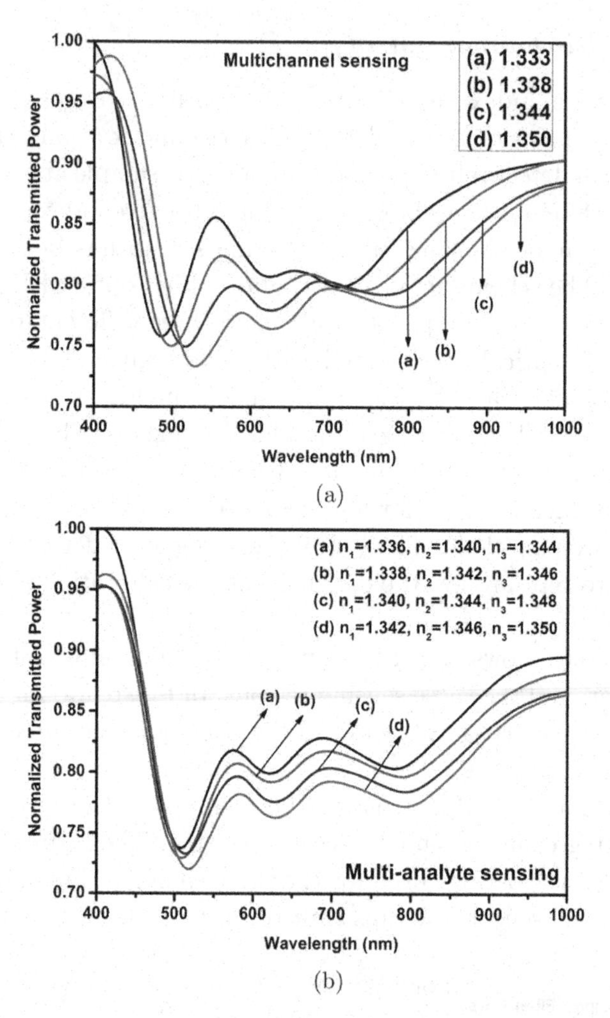

Fig. 6.6. Experimental SPR spectra for samples of (a) same refractive indices, and (b) different refractive indices in all sensing channels. The thicknesses of various films are: silver = 30 nm, gold = 50 nm, copper = 40 nm, and TiO_2 = 10 nm. Reprinted from Ref. [5] with permission from SPIE.

the increasing refractive index of the samples in the case of all the channels. In addition, the dip of channel 1 is the sharpest amongst the three because silver gives sharper SPR curve than gold which has been used for channel 2. The width of the channel 3 is large because of copper layer and that with an additional high index over-layer (TiO_2), which can tune the resonance wavelength and results in the broadening of the SPR curve like silicon. But this is difficult to avoid. The important point for a multi-channel sensor is that all the dips should be well separated because the increase in the separation increases the operating range of each channel. The variation of resonance wavelengths of all the three channels determined from Fig. 6.6(a) for different refractive indices of the sample are plotted in Fig. 6.7. In a small refractive index range the variation is linear. The resonance wavelength regions for all the three channels are different due to the differences in the dielectric properties of the plasmonic metals used.

The simulation of the three channels SPR probe has also been reported using the parameters of the experimental probe. In the case of three channels the reflectance given by Eq. (4.24) becomes

$$R_p = |r_{ps} \cdot r_{pg} \cdot r_{pc}|^2, \tag{6.1}$$

where r_{ps}, r_{pg}, and r_{pc} are the reflection coefficients of silver or Ag, gold or Au, and copper or Cu coated channels, respectively. The method of calculation of transmitted power is the same as described in Chapter 4. The simulated SPR curves are shown in Fig. 6.8(a) and 6.8(b) which are similar to Fig. 6.6(a) and 6.6(b), respectively. The experimental SPR curves and their trends match qualitatively with the simulated ones for multi-channel and multi-analyte sensors. The difference between the two may be due to following reasons: (a) in simulation skew rays have not been considered, (b) no consideration has been given to cylindrical geometry of the fiber in simulations, and (c) in the experiments an un-polarized light rather than p-polarized light has been used.

6.3 pH Sensor

Till now we have discussed about refractive index sensing using SPR technique. However, it is not only the refractive index that can be measured using SPR based fiber optic sensor, but there are many other parameters/analytes that can also be sensed using this kind of sensor. The only requirement is that the refractive

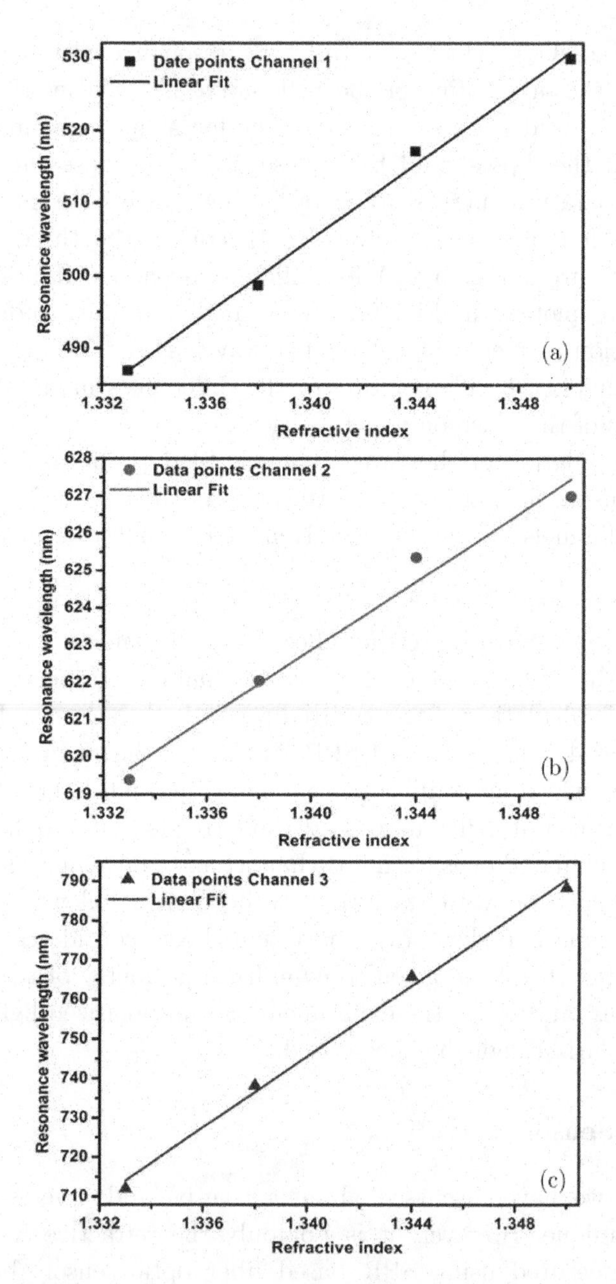

Fig. 6.7. Calibration curves of (a) channel 1, (b) channel 2, and (c) channel 3. Reprinted from Ref. [5] with permission from SPIE.

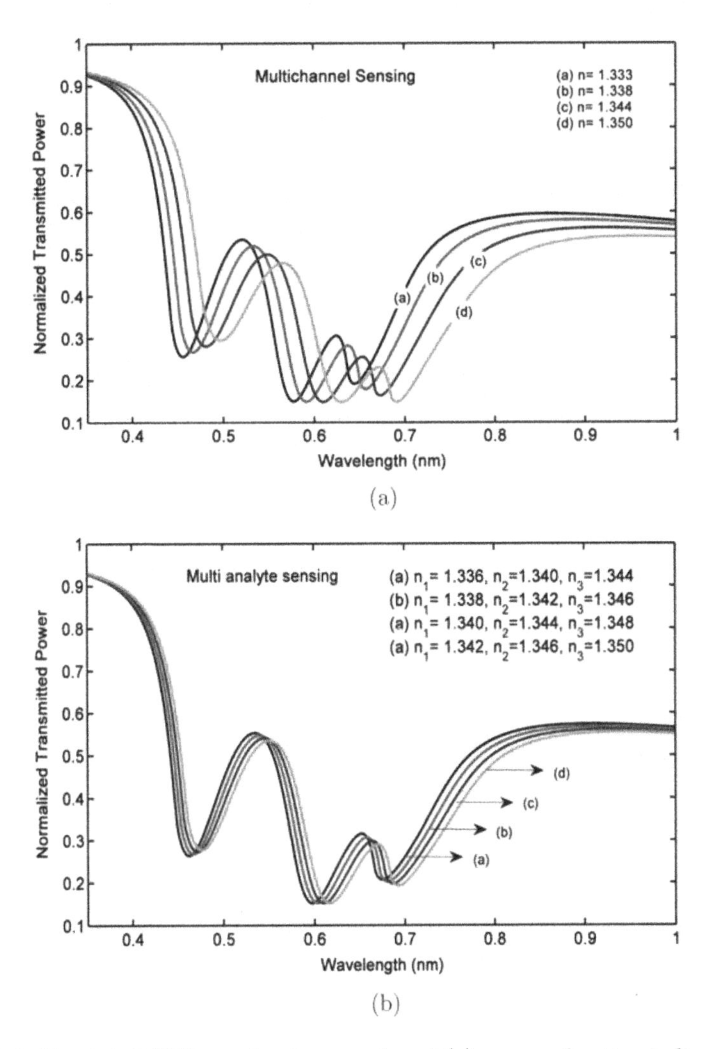

Fig. 6.8. Simulated SPR spectra for samples of (a) same refractive indices, and (b) different refractive indices in all sensing channels. Reprinted from Ref. [5] with permission from SPIE.

index of the medium around the metal layer should change with the change in the quantity/concentration of the parameter/analyte that is to be sensed. One of the parameters that can be measured using this technique other than the refractive index is the pH of a fluid. Its measurement/monitoring is required in various applications such as titration, blood sample, water obtained from different kinds of sources, etc. Generally, pH is measured using a pH-sensitive membrane glass electrode and potentiometric technique. For the measurement of pH, optical fiber sensors have also been proposed. These are based on intensity or wavelength modulation and either use pH-sensitive dyes or long period grating/fiber Bragg grating along with pH-sensitive hydrogel.[6–8] Hydrogel has also been used in other types of pH sensors. In this section, we shall discuss SPR based fiber optic pH sensors utilizing a pH-sensitive hydrogel. Before discussing pH sensor we shall briefly discuss the properties of hydrogel.

Hydrogels are hydrophilic polymers that contain around 30% water in weight. These are sensitive to various chemical and physical stimuli shown in Fig. 6.9. A small change in one of these stimuli present in the fluid surrounding the hydrogel changes the volume of the hydrogel resulting in a change in the refractive index of the hydrogel. This unique property of hydrogel has been widely used in various kinds of studies related to a large number of applications. The hydrophilic nature of the hydrogel is the cause of its swelling or shrinking. Therefore, if a hydrogel layer (specific pH-sensitive hydrogel) is coated over metal layer in a SPR sensor then the pH-dependent swelling/shrinkage of the hydrogel layer can be used for the pH sensing. Here we shall discuss two SPR based fiber optic pH sensors based on this method.

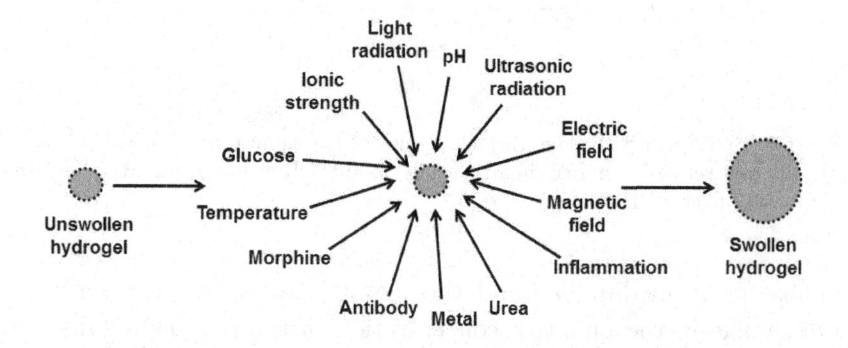

Fig. 6.9. Layout of stimuli action using hydrogel.

6.3.1 *Silver/silicon/hydrogel based pH sensor*

The schematic of the experimental setup for the characterization of a SPR based fiber optic pH sensor fabricated by incorporating a layer of pH sensitive polyacrylamide hydrogel over silver and silicon coated optical fiber is shown in Fig. 6.10.[9] The layer of silicon is added between silver and hydrogel for the enhancement of the sensitivity of the sensor as has been mentioned in Sec. 6.2. The change in pH of an aqueous solution around the probe causes the swelling/shrinkage of hydrogel layer and hence changes the refractive index of the hydrogel layer. According to the principle of SPR based sensing, a change in the refractive index of the dielectric medium interfaced with metal layer shifts the SPR dip and hence the resonance wavelength. To study the sensing capability of the probe, experiments were performed on aqueous solutions of different pH values around the probe. The SPR curves recorded for various pH values at room temperature are shown in Fig. 6.11. A blue shift of SPR curves was observed with the increase in the pH of the fluid. An increase in

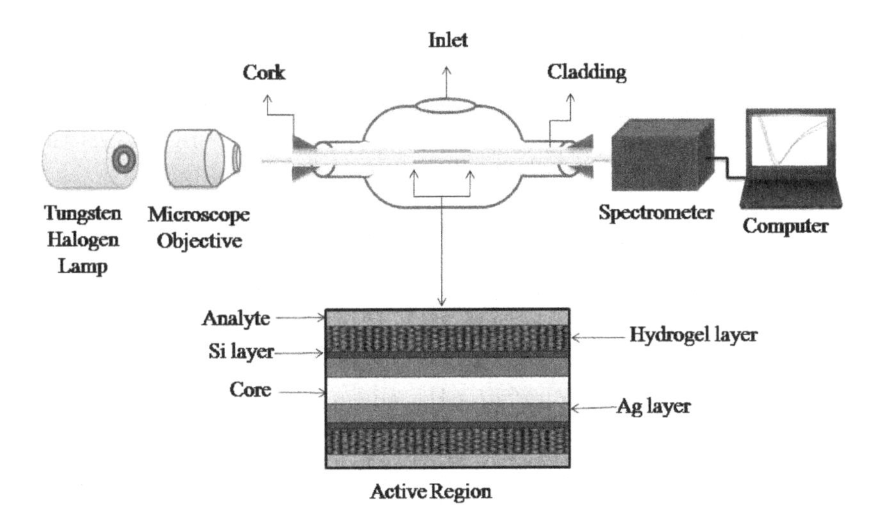

Fig. 6.10. Experimental setup of a SPR based fiber optic pH sensor utilizing silver/silicon/hydrogel layers over unclad core of the fiber. Reprinted from Ref. [9] with permission from Elsevier.

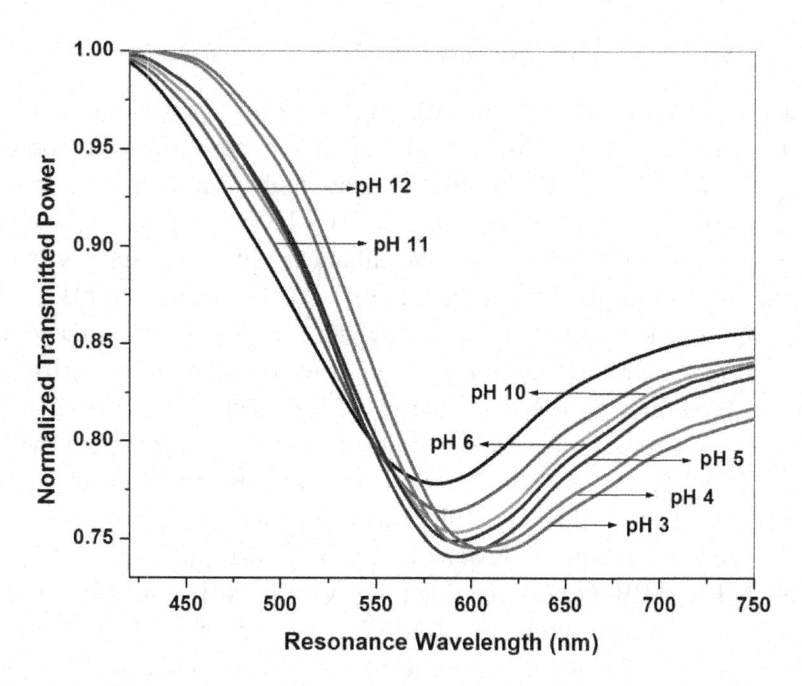

Fig. 6.11. SPR curves at different pH values. Reprinted from Ref. [9] with permission from Elsevier.

the pH causes the swelling of hydrogel under electrostatic repulsion due to the ionization of carboxylic groups. The swelling causes the absorption of water more effectively by the hydrogel, which decreases the refractive index of the hydrogel/sensing layer since the refractive index of the water is lower than that of hydrogel.

The pH variation of the resonance wavelength determined from SPR curves of different pH values of the aqueous samples given in Fig. 6.11 is shown in Fig. 6.12. The calibration curve shows linear variations at lower and higher pH values of the fluid, but in the middle pH range no change in resonance wavelength is observed. This restricts the use of the sensor. In other words, the sensor can only be used for low pH range (3 to 6 pH) and for higher pH range (9 to 12 pH). The sensitivity of the sensor, determined from the slope of the calibration curve, decreases as the pH increases, becomes zero around pH 8.0 and then increases again. The maximum sensitivity obtained is 12.5 nm/pH.

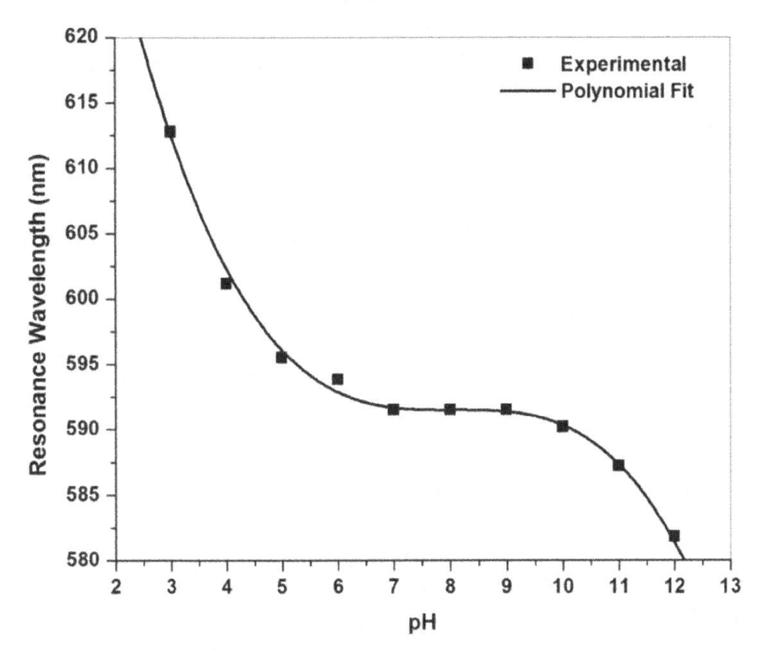

Fig. 6.12. Calibration curve of a SPR based fiber optic pH sensor utilizing silver/silicon/hydrogel layers over unclad core of the fiber. Reprinted from Ref. [9] with permission from Elsevier.

6.3.2 *Silver/indium tin oxide/aluminium/hydrogel based pH sensor*

To further improve the sensitivity of the sensor, coatings on core were changed. The core was coated with silver, indium tin oxide (ITO), aluminium, and then pH-sensitive hydrogel.[10] The SPR spectra for this probe for different pH values of the aqueous sample around the probe are shown in Fig. 6.13. The resonance wavelength, similar to Fig. 6.11, shifts towards blue side. The reason for blue shift is the same as it was for the pH sensor as discussed in Sec. 6.3.1. The aluminium layer in the probe was added to decrease the broadening of SPR curves. The calibration curve of this probe shown in Fig. 6.14 is similar to the pH sensor as discussed in Sec. 6.3.1. The performance of the probe in terms of sensitivity has also been compared with two other probes with following combinations

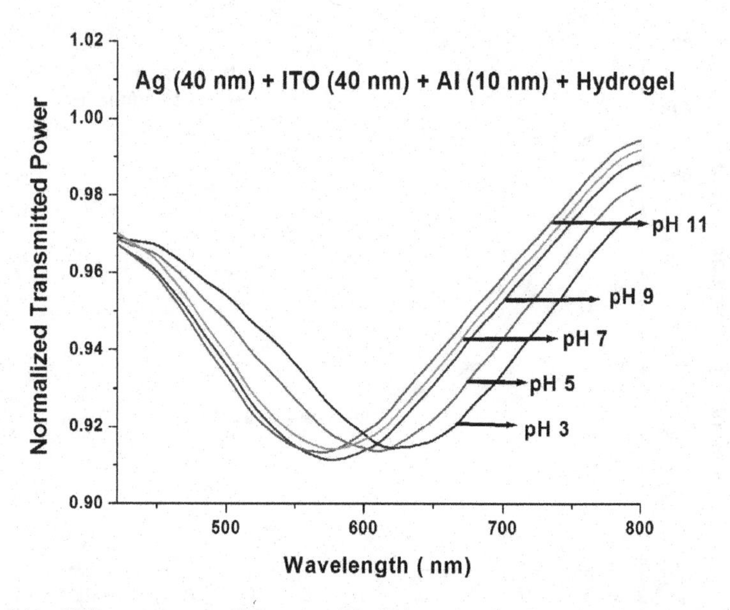

Fig. 6.13. SPR spectra at different pH of the sample. Reprinted from Ref. [10] with permission from Royal Society of Chemistry.

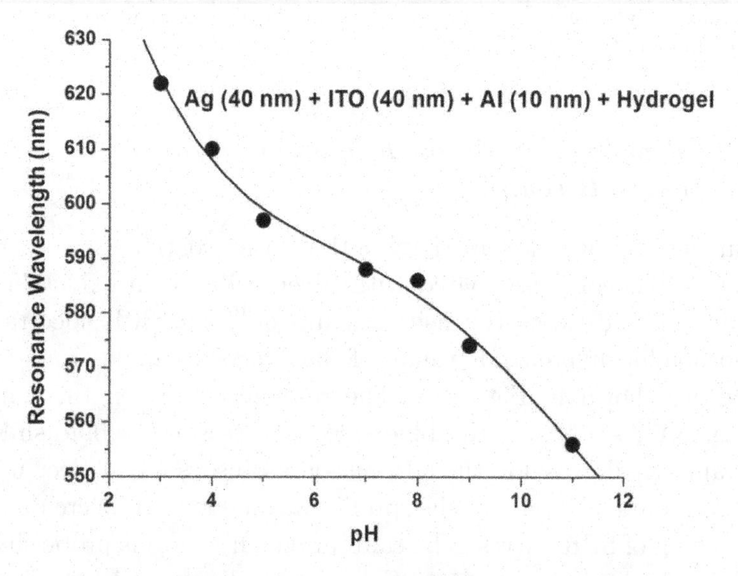

Fig. 6.14. Calibration curve of a SPR based fiber optic pH sensor utilizing silver/ITO/aluminium/hydrogel layers over unclad core of the fiber. Reprinted from Ref. [10] with permission from Royal Society of Chemistry.

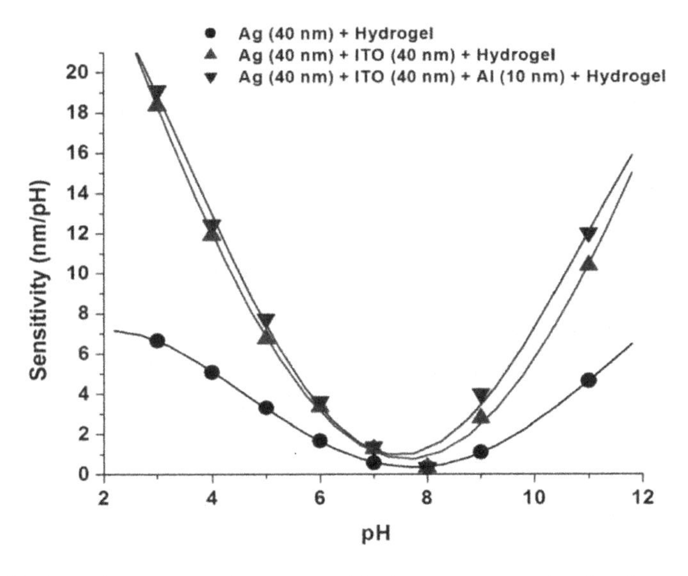

Fig. 6.15. Variation of sensitivity with pH of aqueous samples for three types of the SPR probes. Reprinted from Ref. [10] with permission from Royal Society of Chemistry.

of coatings: (i) silver + hydrogel, and (ii) silver + ITO + hydrogel. The pH variations of sensitivity of all the three probes are shown in Fig. 6.15. The sensitivity is low for the probe with silver and hydrogel layers while it is comparable for the probes with silver/ITO/hydrogel layers and silver/ITO/aluminium/hydrogel layers, the last one having a slightly higher sensitivity. The maximum sensitivity of the sensor is 19.5 nm/pH, which is higher than the sensor as discussed in Sec. 6.3.1.

6.4 Ethanol Sensor

Ethanol is used as a bio-fuel similar to gasoline, diesel, and other fossil fuels. It has got importance due to pollution concerns, limited amount of conventional fuels available, and increasing cost of fuels. The purity of ethanol is the main concern if it is to be used as a bio-fuel. This is because water is completely miscible in ethanol and generally water gets mixed with ethanol during its manufacturing. Therefore, the detection of water content in ethanol is required for it to be used as a bio-fuel or to be used in medicine, pharmaceuticals, and organic chemistry, where its 100% purity is required. Normally, water content in ethanol is measured using Karl Fischer titration but its detection limit is not so good and the process is also slow. Further, it is not suitable for online monitoring and requires sample extraction and chemical treatment. A SPR based fiber optic sensor for the detection of small content of water in ethanol has been reported.[11] The addition of water in ethanol changes its refractive index and hence if kept in contact of gold coated fiber optic SPR probe, the resonance wavelength in the SPR spectrum changes as the content of water changes. Figure 6.16 shows the SPR spectra for ethanol–water solutions with varying percentages of water.[11] It is observed that as the water percentage increases the SPR curve first shifts towards higher wavelength for certain percentage of water and on further increase in water percentage the SPR curve shifts towards the lower wavelength side. This kind of behavior was not expected. The trend should be the same for whole range of water percentage. This is because the refractive index of a homogeneous mixture of two liquids

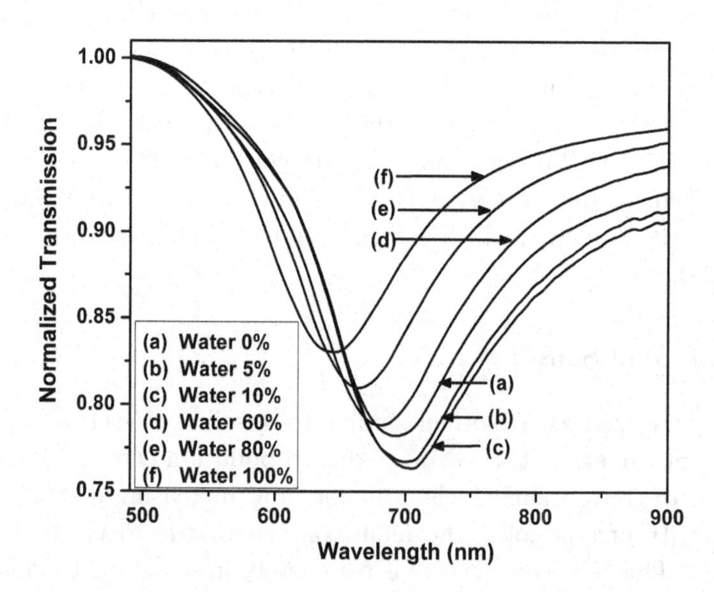

Fig. 6.16. SPR spectra of ethanol-water mixture with varying percentage of water. Reprinted from Ref. [11] with permission from Elsevier.

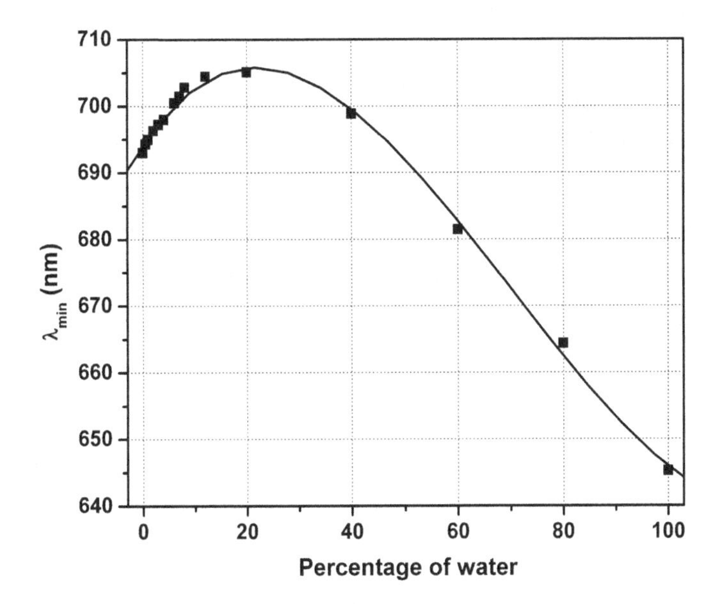

Fig. 6.17. Variation of resonance wavelength with the percentage of water in the mixture. Reprinted from Ref. [11] with permission from Elsevier.

of different refractive indices should lie between the refractive indices of the two individual liquids.

Figure 6.17 shows the calibration curve of the sensor. The resonance wavelength first increases with the increase in the water percentage in ethanol; reaches a maximum value and then decreases with further increase in the water percentage. This kind of behavior suggests that the refractive index variation of ethanol–water mixture with water content does not follow straightforward trend. To test this, the refractive indices of all the ethanol–water samples were measured. The variation of refractive index of the mixture with water percentage is shown in Fig. 6.18. The variation is similar to the variation of resonance wavelength plotted in Fig. 6.17. The maximum refractive index of the mixture is for about 20% of water in ethanol. The similarity between trends of curves in Figs. 6.17 and 6.18 suggests that the results of Fig. 6.17 are not surprising. Such a change in refractive index of the ethanol–water samples is due to the formation of hydrogen (H) bonds between ethanol and water resulting in the formation of ethanol–water clusters at low water

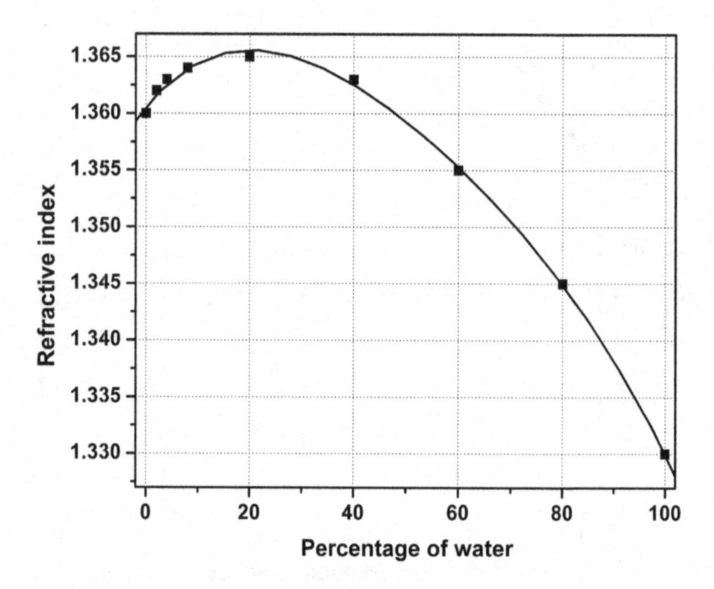

Fig. 6.18. Variation of refractive index with the percentage of water in the mixture. Reprinted from Ref. [11] with permission from Elsevier.

percentage. The formation of clusters increases the density and hence the refractive index of the ethanol–water mixture slightly. Ethanol and water are completely miscible due to the formation of H bonds.

In general, the adulterations in ethanol are important to detect for its practical applications as bio-fuel and in medicine. Therefore, the region of interest of the sensor lies at small water percentages in ethanol–water mixtures. The useful range (0 to 10% in ethanol–water solution) of curve plotted in Fig. 6.17 is shown in Fig. 6.19. The resonance wavelength increases almost linearly with the increase of water percentage in ethanol–water solution. The sensitivity of the sensor reported is about 1.149 nm per water percentage in ethanol.

6.5 Enzyme based Sensors

Enzymes are basically proteins having catalytic activity and are selective towards substrates. These have been used to assay the concentration of different kinds of analytes. Their activity is limited by pH, ionic strength, chemical inhibitors, and temperature. For

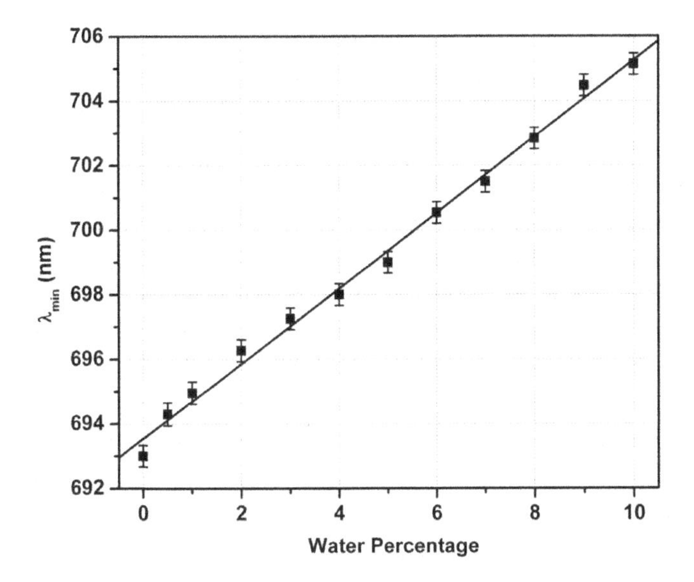

Fig. 6.19. Calibration curve of the sensor. Reprinted from Ref. [11] with permission from Elsevier.

example, some of the enzymes lose their activity when exposed to very high temperatures. In an enzyme based sensor an enzyme is combined with a transducer to produce a signal proportional to the concentration of target analyte. The signal can be due to a change in protein concentration, release or uptake of gases, absorption, reflection or emission of light, change in refractive index of the transducer because of the reaction catalyzed by the enzyme. Due to specificity and catalytic properties of enzymes, they have been widely used in biosensors. The enzymes are immobilized at the surface of the transducer by adsorption, covalent attachment, or entrapment in a gel. These are commercially available in high purity which makes them very attractive for the mass production of enzyme based sensors. In this section, we shall discuss enzyme based fiber optic SPR sensors for the detection of various analytes.

6.5.1 *Urea sensor*

Urea is a waste product of many living organisms, and is the major organic component of human urine. It is produced by the liver and transported to kidneys via blood for excretion. An adult typically excretes about 25 grams of urea per day. Elevated levels of urea or change in urea concentration in blood are clinically significant in many renal disorders such as kidney and liver dis-functioning. Also any condition which impairs the elimination of urea by the kidneys can be fatal. To reverse the condition, either the cause of the kidney failure must be removed, or the patient must undergo blood-dialysis to remove the wastes from the blood. Thus, the estimation of urea concentration is very important for the diagnosis of such diseases. Various kinds of biosensors have been reported in the literature for the determination of the level of urea. However, the optical biosensors are more advantageous for biomedical applications because they do not need a reference cell/electrode and are not affected by the electromagnetic interference. Out of the many optical techniques available, SPR technique has got tremendous attention for detection of biological and chemical analytes including urea. Here we shall discuss a SPR based fiber optic urea sensor that has been reported in the literature.[12] For this sensor the probe was prepared by coating layers of silver, silicon and the enzyme, urease, over the unclad core of an optical fiber. The enzyme, urease, was immobilized using gel entrapment technique. The details of gel entrapment method are discussed in Sec. 5.2.2 of Chapter 5. To characterize the probe, the samples of various urea concentrations were placed in the flow cell attached with the probe. The samples of urea were prepared in sodium phosphate buffer solution. Figure 6.20 shows the SPR spectra for various concentrations of urea. The SPR curve shifts towards lower wavelength as the concentration of urea increases. The variation of resonance wavelength with the concentration of urea is shown in Fig. 6.21. The resonance wavelength decreases with the increase in the concentration of urea, which is due to the decrease in the refractive index of the enzyme immobilized layer because of the formation of the complex between urease and urea. The variation is non-linear and the resonance wavelength appears to saturate at higher concentration of urea due to the limited quantity of enzyme in the film. The disadvantage of the sensor is its limited operating range which can be increased if the quantity of enzyme in the film is

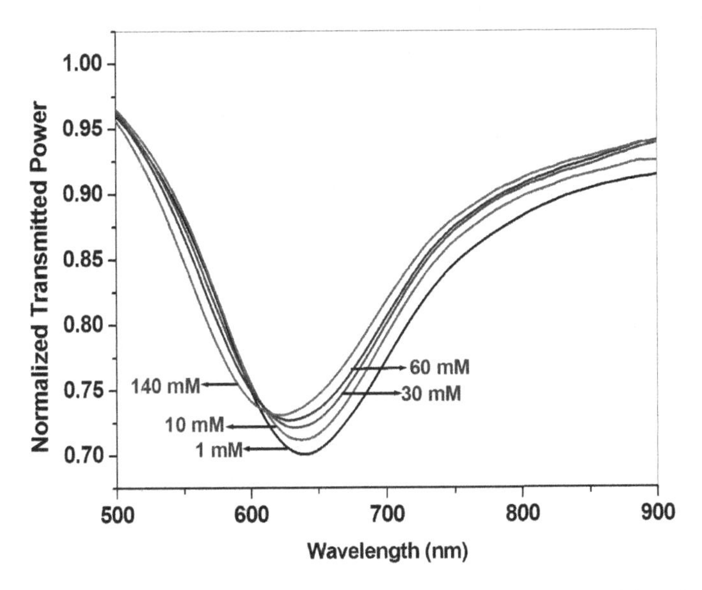

Fig. 6.20. SPR curves for different concentrations of urea. Reprinted from Ref. [12] with permission from Elsevier.

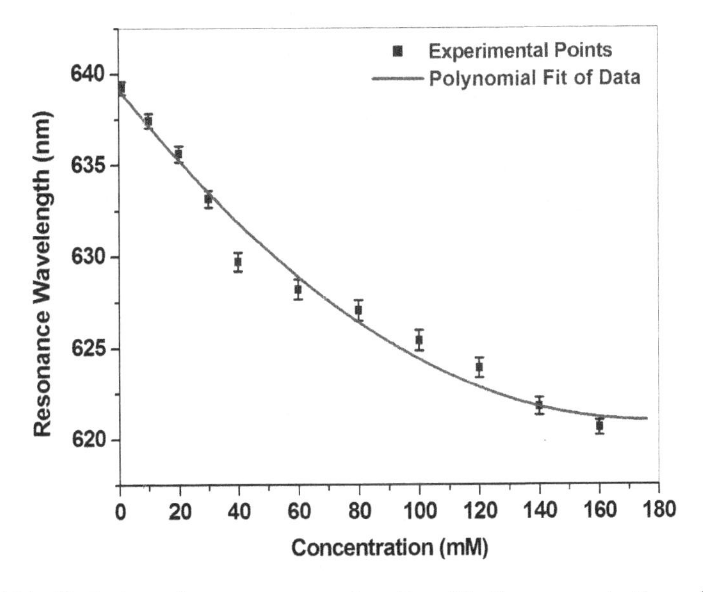

Fig. 6.21. Variation of resonance wavelength with the concentration of urea. Reprinted from Ref. [12] with permission from Elsevier.

increased. However, the quantity of enzyme can be increased only up to a certain amount otherwise the gel cannot hold it within the film.

In the urea probe silicon layer was used due to the following three reasons: (i) protection of silver from oxidation, (ii) enhancement of the sensitivity of the sensor, and (iii) protection of the activity of enzyme in gel from deactivation by the metal layer.

6.5.2 *Naringin sensor*

Sometimes during processing or after that, the citrus fruit juice becomes bitter which affects its consumer acceptability. The bitterness in fruit juice is caused by naringin whose content in the juice varies from 0.02% to 0.07%. A fiber optic sensor based on SPR technique and gel entrapment method for the immobilization of enzyme naringinase over silver coated core of the fiber was reported for the detection of naringin.[13] The method of testing of the probe was the same as that of urea discussed in the previous section. The resonance wavelengths were determined from the SPR spectra of fluids containing different amounts of naringin placed around the probe. Contrary to urea sensor, the resonance wavelength was found to increase with the increase in the concentration of naringin, which implied that the refractive index of gel layer increases as the concentration of naringin increases. Further, the sensitivity of the sensor increased with the increase in the concentration of naringin implying that the enzyme was not completely used for the maximum quantity of naringin used in the study.

6.5.3 *Organophosphate pesticide*

SPR based fiber optic sensor utilizing enzymatic reaction has been reported for the detection of organophosphate pesticide, chlorphyrifos.[14] The enzyme used for immobilization over silver coated core and for the sensing was acetylcholinesterase (AChE). However, the method of detection was different from that used for the sensing of urea and naringin. In the case of urea and naringin sensors, the analyte binds with the enzyme, while in the case of pesticide sensing, it acts as an inhibitor to the substrate (acetlythiocholine iodide) present in fluid sample and binds to the enzyme AChE; a method called competitive binding. In this case, for the fixed concentration of the substrate in the fluid, the resonance wavelength decreases with the increase in the concentration of the pesticide. Further, the sensitivity was reported to decrease with the increase in the concentration of the pesticide in the fluid around the probe.

6.5.4 *Phenolic compounds*

Phenolic compounds are the polluting chemicals widely distributed throughout the environment. These are highly toxic and their persistence in the environment is a potential hazard to living organism if absorbed. There are various standard procedures/techniques available to monitor the concentration of phenolic compounds. Some of the techniques/methods are expensive, time consuming, and require trained experts. A SPR based fiber optic phenol biosensor has been recently reported.[15] The sensing probe was prepared by depositing enzyme, tryosinase, over the silver coated unclad core of an optical fiber using gel entrapment technique. The gel layer was of polyacrylamide. The detection of following four functional groups attached to the phenolic compounds was studied: phenol, catechol, *m*-cresol, and 4-chlorophenol.

The SPR spectra recorded for $0\,\mu$M to $1000\,\mu$M concentration of catechol solutions are shown in Fig. 6.22. The figure shows a red shift of the SPR dip as the concentration of catechol increases. The shift occurs due to the enzymatic reaction between substrate (phenolic compound) and enzyme forming a complex called enzyme–substrate complex. The breakdown of the complex releases free enzyme and changes the effective refractive index of the gel layer resulting in the red shift of the resonance wavelength of SPR spectra. This can also be explained by the following enzymatic reaction catalyzed by the enzyme tyrosinase and the refractive indices of the gel layer and the by-products:

$$\text{Phenol} + \text{Tyrosinase } (O_2) \rightarrow \text{Catechol}$$

$$\text{Catechol} + \text{Tyrosinase } (O_2) \rightarrow \text{o-Quinone} + H_2O$$

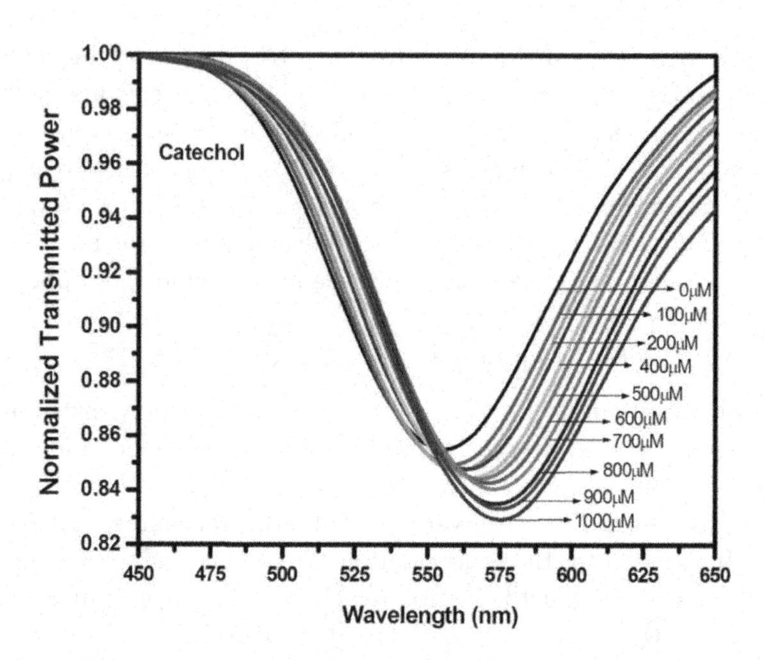

Fig. 6.22. SPR spectra of catechol solutions of different concentrations prepared in air-saturated 0.1 M and pH 7.0 phosphate buffer solutions. Reprinted from Ref. [15] with permission from Elsevier.

The refractive index of polyacrylamide gel layer (1.47) is small as compared to the refractive indices of the by-products, catechol (1.61298) and o-quinone (1.5432), which form during enzymatic reaction. Since the refractive index of the gel layer is lower than the refractive indices of by-products, the effective refractive index of the gel layer increases as the concentration of phenol compound increases. In addition, the by-product, o-quinone, is unstable and hence undergoes a non-enzymatic polymerization to form pigments called melanin which has even higher refractive index (1.7). This further increases the effective refractive index of gel layer. Thus, the effective refractive index of gel layer increases as the concentration of phenolic compounds in the fluid around the probe increases resulting in the shift in the resonance wavelength towards red side.

The variation of resonance wavelength with the concentration of various phenolic compounds (catechol, m-cresol, 4-chlorophenol, and phenol) is shown in Fig. 6.23. The variation, in the case of all phenolic

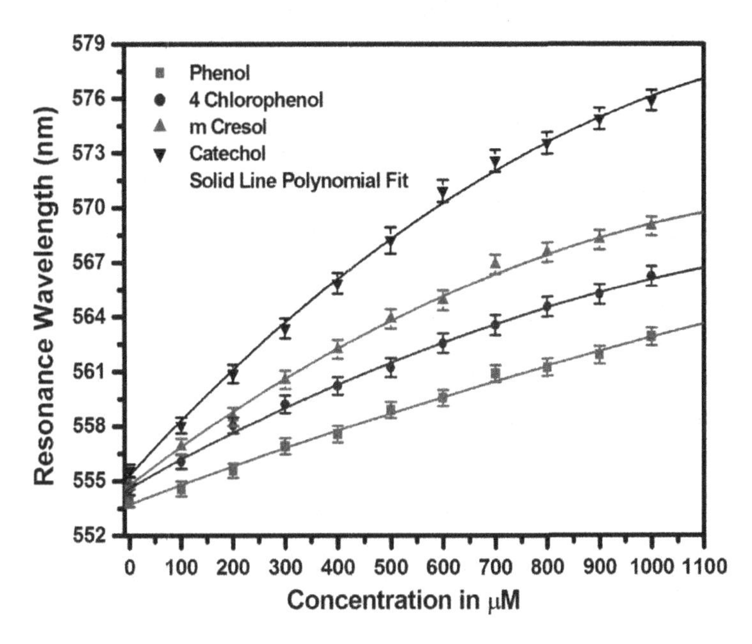

Fig. 6.23. Variation of resonance wavelength with the concentration for four phenolic compounds. Reprinted from Ref. [15] with permission from Elsevier.

compounds, is non-linear and saturates at higher concentrations. As mentioned earlier, the saturation, in this case also, appears to be due to the limited quantity of the enzyme in the gel film. The gel layer possesses a fixed number of enzyme active sites and these sites of the enzyme get occupied when a particular concentration of the substrate (analyte) is present in the aqueous sample. Therefore, further increase in the concentration of the substrate in the sample does not change the dielectric properties of the gel layer because all the sites are already occupied. The sensitivity (S) of the probe determined from the slope of the curve shows that it gradually decreases as the concentration of phenolic compound increases for all the phenolic compounds considered. For a fixed concentration of phenolic compound, sensitivities of different phenolic compounds are in the following order: $S_{\text{catechol}} > S_{\text{m-cresol}} > S_{\text{chlorophenol}} > S_{\text{phenol}}$. The difference in sensitivity between phenolic compounds might be due to the conformational effects and diffusional capability of phenolic compounds to the immobilization matrix. This means that the hydrophobic/hydrophilic characteristics of the immobilized layer play an important role in the sensitivity. The higher sensitivity of the sensor for catechol (diphenolic) is due to more hydrophilic nature of the gel layer for it as compared to other phenolic compounds. Since the polyacrylamide gel is hydrophilic it increases diffusion of catechol molecules through immobilization matrix, which increases the binding of catechol to the enzyme. The difference in sensitivity between mono-phenolic compounds may be due to the following reasons: (i) the analytes are different in analyte–enzyme reaction, (ii) the solubility of phenolic compounds in the immobilization matrix may be different, and (iii) the enzymatic reaction is affected by various parameters such as the acidity constant of the phenolic compound and acceptor/donor effects which provides the binding between carbon and oxygen of tyrosinase. However, response time is nearly the same for o-diphenolic and mono-phenolic compounds.

6.5.5 *Glucose sensor*

Diabetes is one of the chronic diseases to which many people are affected these days. Change in the diets and use of medication and insulin allow patients to live normal lives. However, the monitoring and maintenance of glucose levels is very important to save the patients from complications associated with diabetes. Many methods have been used to measure the glucose content for diagnostic purpose. A fiber optic sensor based on SPR technique and enzymatic reaction for the detection of low concentration of glucose has been recently reported.[16] The SPR technique with glucose specific enzyme, glucose oxidase, is very specific and sensitive to glucose concentration. Its performance is not affected by the other constituents found in blood. The fabrication of the glucose probe is similar to the urea probe as discussed earlier. It consists of coatings of silver and silicon layers followed by immobilization of glucose specific enzyme (glucose oxidase) using gel entrapment method over unclad core of the fiber.

The SPR spectra recorded for different concentrations of glucose in pH 7.0 sodium phosphate buffer solution around the probe are given in Fig. 6.24. The SPR curve (and the resonance wavelength)

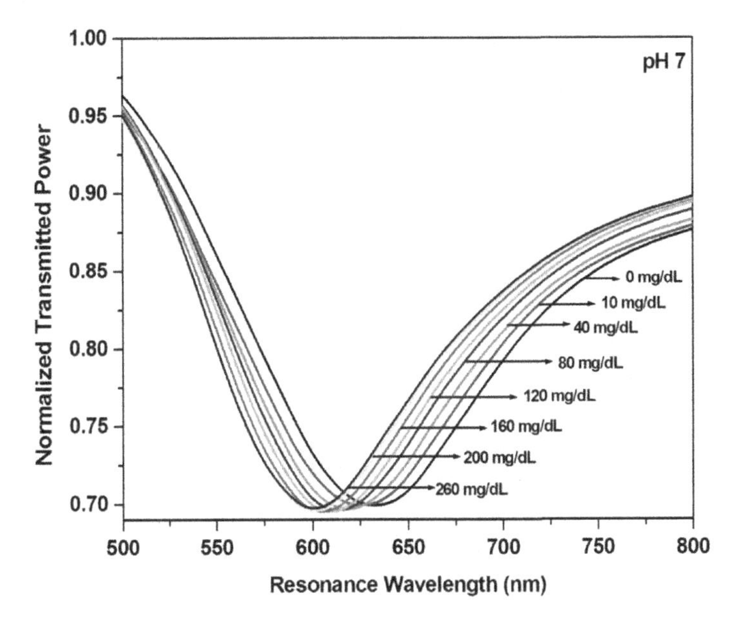

Fig. 6.24. SPR spectra of glucose solutions of different concentrations prepared in pH 7.0 phosphate buffer solutions. Reprinted from Ref. [16] with permission from Elsevier.

shifts towards lower wavelength as the concentration of glucose increases. It is also known that the resonance wavelength decreases only when the refractive index of the sensing medium around the probe decreases. Thus, the refractive index of the gel layer is decreasing as the concentration of glucose is increasing. This can be explained on the basis of the following reaction taking place in the gel layer catalyzed by the enzyme glucose oxidase:

$$\text{Glucose} + O_2 \xrightarrow{\text{Glucose oxidase}} \text{Gluconic acid} + H_2O_2$$

According to this reaction the conversion of glucose to gluconic acid and hydrogen peroxide (H_2O_2) occurs in the presence of glucose oxidase enzyme. The decrease in the resonance wavelength and hence the refractive index of the gel layer can also be explained on the basis of the refractive indices of polyacrylamide gel layer and the enzymatic reaction products. The refractive index of polyacrylamide gel layer

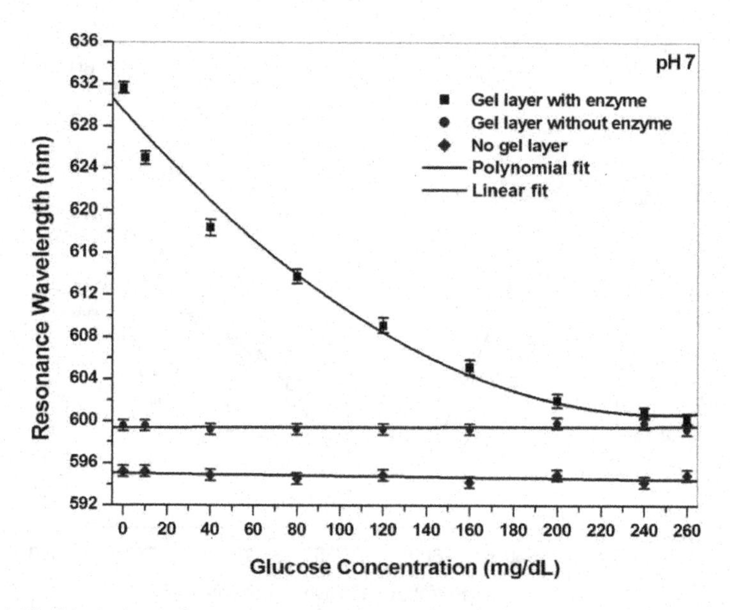

Fig. 6.25. Variation of resonance wavelength with the concentration of glucose prepared in buffer solution of pH 7.0 for three probes with different kinds of layers. Reprinted from Ref. [16] with permission from Elsevier.

is 1.47 while the refractive indices of the products, gluconic acid, and H_2O_2, are 1.4161 and 1.414, respectively. Since the refractive indices of products are lower than the refractive index of gel layer, the effective refractive index of the gel layer decreases as the concentration of the glucose in the sample is increased.

The variation of resonance wavelength with the concentration of glucose is shown in Fig. 6.25. The resonance wavelength decreases non-linearly with the concentration of glucose and saturates at higher concentration which is due to the limited quantity of enzyme entrapped in the gel layer. To confirm that the shift in resonance wavelength is due to enzymatic reaction, control experiments were performed on the probes fabricated with the following coatings: (i) silver–silicon layers, and (ii) silver–silicon–gel layers without enzyme. The variations of resonance wavelength with glucose concentration of these probes also shown in Fig. 6.25 do not show any appreciable shift in resonance wavelength which means that the shift in resonance wavelength is solely due to the enzymatic reaction.

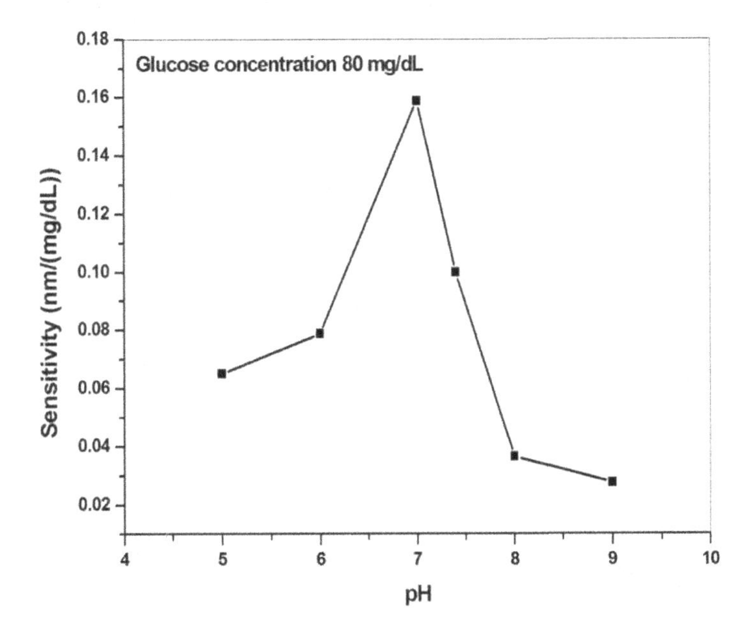

Fig. 6.26. Variation of sensitivity of the sensor with pH of the sample. Reprinted from Ref. [16] with permission from Elsevier.

For all these measurements the solutions of glucose with different concentrations were prepared in buffer solution of pH 7.0. Since the activity of an enzyme depends on the pH, the performance of the glucose sensor may be affected by the pH of the sample fluid. Figure 6.26 shows the pH dependence of the sensitivity of the glucose sensor. The maximum sensitivity was obtained for pH 7.0 which may be due to the maximum reactivity of the enzyme, glucose oxidase, at pH 7.0.

Although the optimum working pH of the sensor is 7.0 but it is close to blood pH which is 7.4. Further, the operating range of the sensor is from 0–260 mg/dL glucose concentration and falls in the physiological range of human blood glucose, therefore, the sensor may find its application in biomedical sciences.

6.5.6 *Low density lipoprotein sensor*

Cholesterol is a small molecule whose level in human blood lies in the range 20 mg/dL to 200 mg/dL. For a healthy person, its level in blood must lie within a certain concentration range. The total blood cholesterol consists of four constituent lipoproteins namely very low density lipoprotein (VLDL), intermediate density lipoprotein (IDL), low density lipoprotein (LDL), and high density lipoprotein (HDL). Among these lipoproteins, LDL, is one of the fatty acids and cholesterol rich lipoprotein that transports cholesterol into artery walls. The LDL molecules bind with LDL receptor apolipoprotien B-100 (AAB) during the transportation of cholesterol into artery resulting in its accumulation on artery walls and consequently forming atherosclerotic plaques. These plaques are the main cause of heart diseases. Being harmful, LDL is called bad cholesterol. In terms of its effects on heart diseases, the level of LDL $\leq$100 mg/dL is optimal. Therefore, regular monitoring and control of LDL in human blood is required. A fiber optic sensor for the detection of LDL using SPR technique has been reported.[17] The probe consists of layers of gold, self-assembled monolayers (SAMs) of 4-ATP and antiapolipoprotein B (AAB). A schematic of step by step formation of self-assembled monolayers over the gold film coated optical fiber is given in Fig. 6.27. AAB is a very good biomolecular recognition element (BRE) for binding of LDL. Before the covalent immobilization of AAB over 4-ATP/Au surface, AAB was first conjugated with N-ethyl-N-(3-dimethylaminopropyl carbodimide) (EDC) and N- hydroxysuccinimide (NHS). EDC is a widely used cross-linker of negligible length and is frequently used for immobilization of proteins over various surfaces. The conjugation of AAB with EDC–NHS allows the direct immobilization of AAB on 4-ATP/Au surface.

The SPR spectra of the probe for different concentrations of LDL in PBS show a red shift with the increase in the concentration of LDL in the LDL concentration range 0 mg/dL to 190 mg/dL. The shift occurs due to the binding of LDL molecules over AAB ligands which change the effective refractive index of the sensing surface. The sensitivity of the sensor is reported to be 0.18 nm/(mg/dL).[17] The sensor's response time is much smaller than that of the currently available sensors.

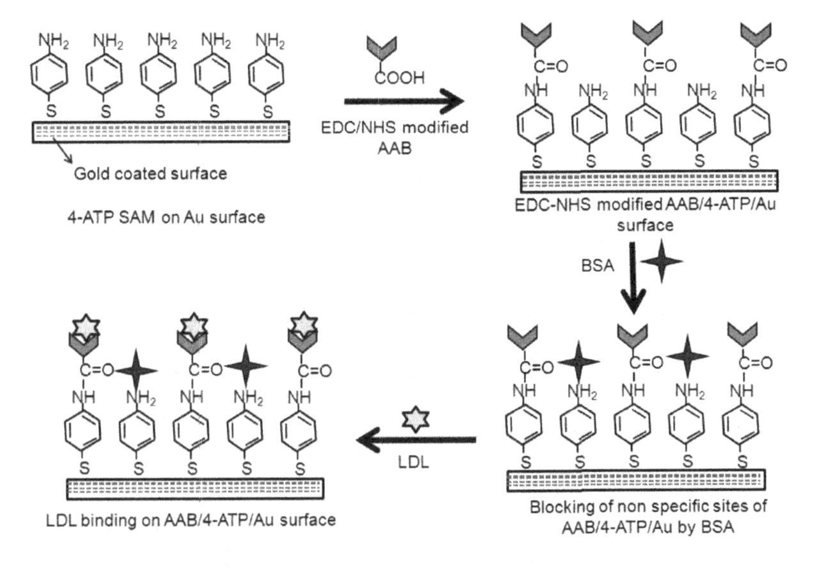

Fig. 6.27. Schematic of stepwise preparation of Au/4-ATP/AAB modified fiber surface for binding of LDL. Adapted from Ref. [17].

6.6 Molecular Imprinting based Sensors

Inverse colloidal crystal templating and molecular imprinting (MIP) are some of the techniques extensively used for the fabrication of various chemical and bio-sensors. In some cases a combination of inverse colloidal crystal templating and MIP techniques along with hydrogel has been used for the detection of variety of chemical analytes. Hydrogel is a part of polymerization process in the preparation of molecular imprinted templates. Hydrogel, as discussed in Sec. 6.3, swells/shrinks when exposed to particular chemical/biological stimuli and gives rise to volume change and hence the change in the refractive index. Inverse colloidal crystal template is a structure of long range ordering of particles which results in a high surface to volume ratio. It is generally formed by packing uniform monodisperse polystyrene nanospheres into 3-dimensional array with interstitial space filled by a fluid and then left to dry. The procedure is followed by the removal of nanospheres to leave the new surface with voids which preserve the most valuable property of the inverse colloidal crystal. MIP is a technique for the creation of binding sites on the sensing layer with the memory of size, shape, and functional group of imprinted molecules (target molecules). In this section we shall discuss two sensors, one utilizing MIP, while the other uses combination of inverse colloidal crystal template and MIP.

6.6.1 *Vitamin B_3 sensor*

Vitamin B_3, also known as 3-pyrydinecarboxamide (PA), is required for cell respiration and release of energy and metabolism of carbohydrates, fat, and proteins in human body. It is responsible for keeping skin healthy and proper functioning of the nervous system. There is a recommended dietary allowance (RDA) of vitamin B_3 for different age groups for healthy functioning of human body. Its intake is through food and some time in pharmaceutical forms also. For the fabrication of vitamin B_3 sensing probe, a combination of inverse colloidal crystal template and MIP was used.

The inverse colloidal crystal template, on silver coated fiber optic probe, was prepared by dipping the probe in aqueous solution of monodisperse polystyrene nanospheres using a vertical deposition–evaporation method.[18] A complete evaporation of water from the aqueous polystyrene nanospheres solution gave colloidal crystal templated optical fiber probe on which the polymerization medium was poured drop by drop along the sensing region to infiltrate the colloidal crystal template. This polymerization medium had a particular ratio of monomer acrylamide, crosslinker N, N-methylenebisacrylamide (BIS), phosphate buffer, and Millipore water. After polymerization the colloidal crystal template probe was incubated in dimethylbenzene for complete removal of colloidal crystal. The probe was further incubated in aqueous acetic acid to completely remove the embedded PA molecules. The removal of imprinted PA molecules left many specific binding sites possessing same shape and size functional groups, which were capable of recognizing PA molecules. In other words, the surface had many recognizing nano-cavities also called binding sites. The sensing probe was then named as inverse colloidal crystal templated-PA imprinted optical fiber probe. In this probe, certain specific recognizing nano-cavities for PA molecules were formed.

For the characterization, the SPR spectra were recorded for different concentrations of solutions of PA.[18] These SPR spectra of PA samples of different concentrations varying from 0 mg/ml to 10 mg/ml are shown in Fig. 6.28(a). The SPR spectra show a red shift with increasing concentration of PA. The reason of shift in resonance wavelength is that when the sensing surface is exposed to PA sample solution, the PA molecules interact with binding sites through non-covalent interactions. Due to these interactions, PA molecules get stuck in nano-cavities resulting in the change in the morphology of the polymerization matrix of the sensing surface. These bindings of PA molecules to nano-cavities change the effective refractive index of the sensing surface. The change in the effective refractive index causes the red shift of the resonance wavelength of SPR spectra. Figure 6.28(b) shows the variation of resonance wavelength with PA concentration. It is found that the resonance wavelength increases with the increase in the concentration of PA in sample solution. The total shift in resonance wavelength obtained is 16.651 nm for the change in concentration from 0 mg/ml to 10 mg/ml. The sensitivity of the sensor calculated from the slope of the linear curve is 1.483 nm/(mg/ml).

6.6.2 *Tetracycline sensor*

For the detection of vitamin B_3, inverse colloidal crystal template with MIP was used. Here we shall discuss the detection of tetracycline (TC) using MIP only. TC is one of the antibiotics used for the treatment of the bacterial infection including pneumonia, acne, and infections of skin. It falls in a class of medication called TC antibiotics. As we know that the consumption of antibiotics in large amount is not recommended because it may create difficulties associated with allergic symptoms and skin disease. The consumption of large quantity of TC with food is dangerous for human being. The maximum and safe residue limit of TC in different foods is fixed, for

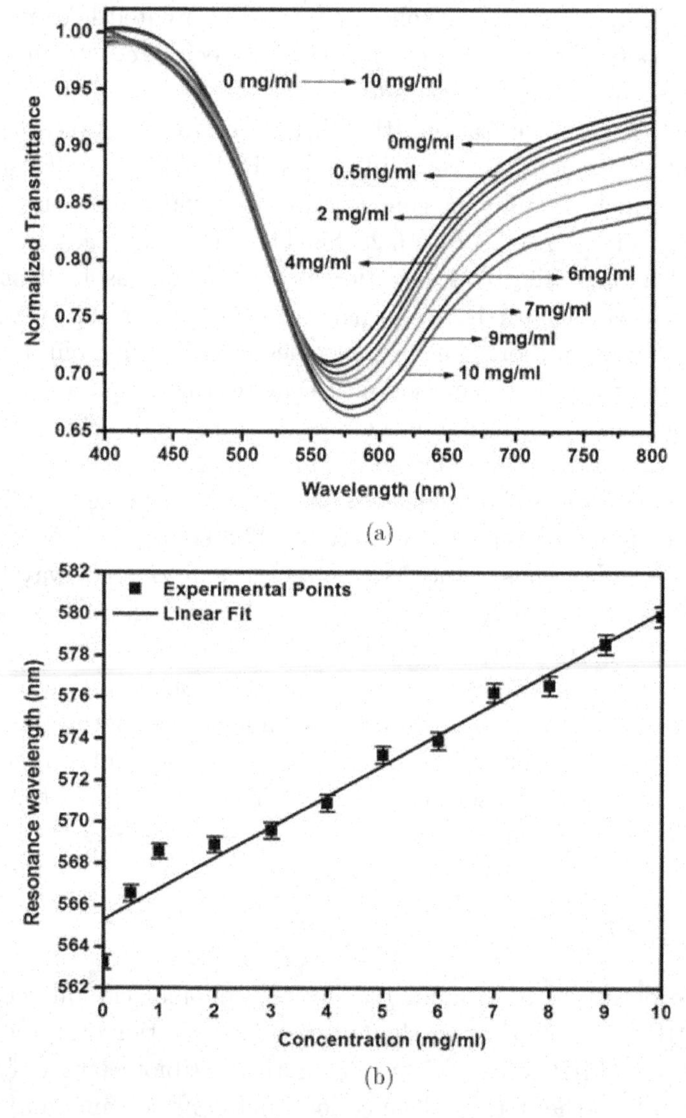

Fig. 6.28. (a) SPR spectra for different concentration of PA, and (b) variation of resonance wavelength with concentration of PA for colloidal crystal templated-PA imprinted optical fiber SPR probe. Reprinted from Ref. [18] with permission from Elsevier.

example, 0.1 mg/kg in milk and 0.01 mg/kg in honey. Therefore, the detection of the concentration of TC in food stuff is an important issue of concern to control food safety. A SPR based fiber optic TC sensor using MIP technique for the preparation of the sensing surface has been reported.[19] The sensing surface prepared using MIP generates specific binding sites of TC (target molecules). When TC molecules bind on sensing surface, the morphology of the surface changes, thus, changing its refractive index and hence SPR technique is useful for sensing of TC. The MIP surface is stable and is very selective of the target molecule. The polymer matrix attached on the silver coated fiber core remains there even after the removal of TC molecules and creating nano-cavities on the MIP matrix. The TC–MIP probe has several binding sites/nano-cavities and functional groups for the TC molecules.

The SPR spectra for tetracycline samples of different concentrations ranging from 0.0 μM to 0.96 μM prepared in de-ionized water are shown in Fig. 6.29(a). The SPR curves shift toward the higher wavelength with the increasing concentration of TC. The shift in SPR curves occurs due to the binding of TC molecules in the nano-cavities created on the TC–MIP matrix when the sample of TC comes in the vicinity of the sensing surface. The binding causes change in the volume of the MIP matrix, resulting in the change in the refractive index of the sensing surface and hence the red shift of the SPR curve.

To check the selectivity of the TC–MIP probe, SPR curves were recorded for the oxy-tetracycline (OTC) sample solutions of concentration range 0.0 μM to 0.96 μM over TC–MIP probe. The SPR spectra for OTC samples are shown in Fig. 6.29(b). As expected, all SPR curves almost overlap with each other implying that OTC molecules are not compatible with nano-cavities on TC–MIP probe. In other words, no binding occurs between OTC molecules and polymer matrix of TC–MIP. Hence, due to the shape, size, and functional group of binding sites on MIP matrix, only specific target molecules can bind to these, causing the change in the volume and morphology of the sensing surface which results in the shift of SPR curves. The MIP–matrix was created by polymerization over metal coated optical fiber probe with polymer medium. This consists of

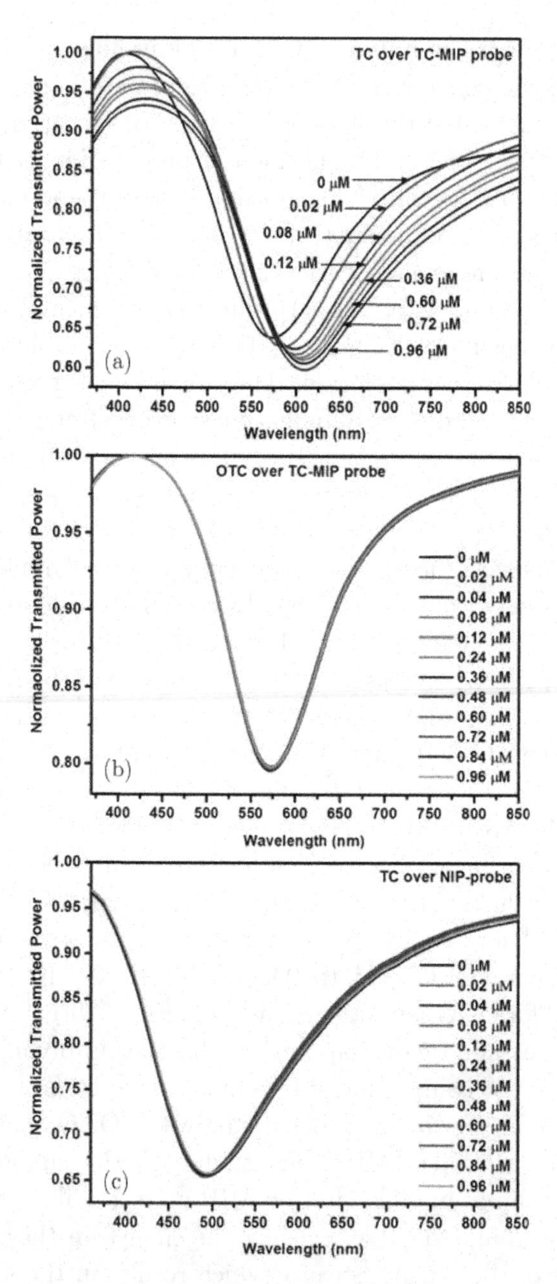

Fig. 6.29. SPR spectra for different concentrations of (a) TC over TC–MIP probe, (b) OTC over TC–MIP probe and (c) TC over NIP probe. Reprinted from Ref. [19] with permission from Royal Society of Chemistry.

monomers, cross-linker and the target molecule (TC). Polymerization is followed by the extraction of target molecules. The probe prepared by not adding target molecules in polymerization solution also does not show any shift of SPR curves. This is shown in Fig. 6.29(c) for all TC samples over non-molecularly imprinted (NIP) probe. This is because there is no interaction between TC molecules and polymerization matrix because no binding sites/nano-cavities are formed in NIP probe as target molecules were not added in the polymerization medium.

The variation of shift in resonance wavelength with the concentration of TCs and OTCs in all the three cases are shown in Fig. 6.30. The shift in resonance wavelength for TC samples over TC–MIP probe increases rapidly for low concentrations of TC samples, but after that it increases slowly and appears to saturate at higher concentrations of TC samples. The total shift in resonance wavelength is 35.892 nm when TC–MIP probe is dipped in TC samples for 0.0 μM

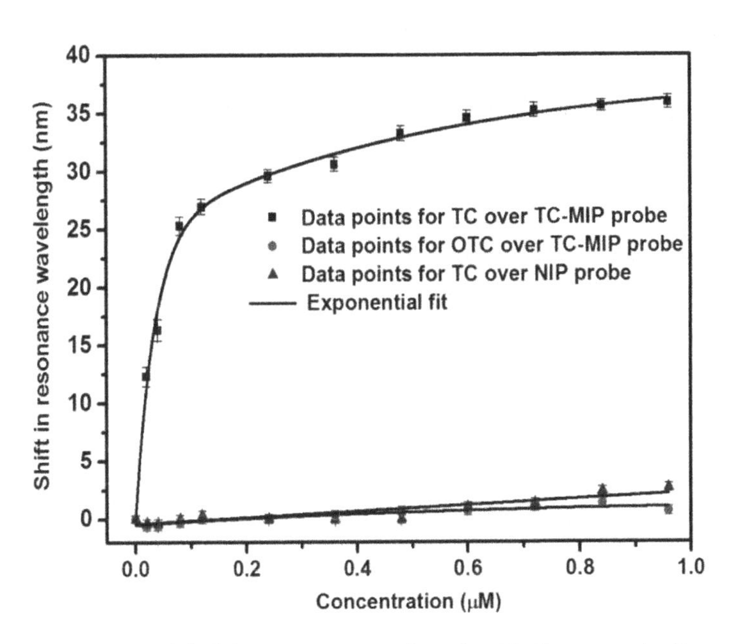

Fig. 6.30. Variation of shift in resonance wavelength with the concentrations of TC and OTC samples. Reprinted from Ref. [19] with permission from Royal Society of Chemistry.

to 0.96 μM concentration range. From all the figures and discussion, it was concluded that the nano-cavities in TC–MIP probe operate only for TC samples but not for OTC samples while NIP probe does not respond to TC samples. It is a good support for the selectivity of the MIP and recognition capabilities of the target molecules.

The same procedure was followed for the OTC–MIP probe fabricated adding OTC in master solution to confirm the role of imprinting. The SPR spectra for different concentrations of OTC samples over OTC–MIP probe are shown in Fig. 6.31. Similar to TC–MIP probe, the SPR curves shift towards the higher wavelength with the increasing concentration of OTC samples and binding of OTC. The selectivity of OTC–MIP probe was checked using TC samples over OTC–MIP probe. Further, NIP probe could not show any shift in SPR curves due to the absence of binding sites on its surface as in the case of TC samples. The variation of shift in resonance wavelength determined using resonance wavelengths of the SPR curves as a function of concentration of the sample is shown in Fig. 6.32 for the three cases. The total shift in resonance wavelength for OTC sample is 14.668 nm for the concentration range 0.0 μM to 0.96 μM. Similar to TC–MIP probe, the resonance wavelength shift is high for small concentration of OTC and appears to saturate for high concentrations of OTC.

The sensitivities of the TC–MIP and OTC–MIP probes as a function of concentration of respective analyte are shown in Fig. 6.33. In both the cases, the sensitivity decreases with the increasing concentrations of respective analyte. For a given concentration of analyte the sensitivity of the TC–MIP probe is greater than the sensitivity of the OTC–MIP probe. This may be due to the interaction of nano-cavities with corresponding target molecules having different functional groups and the change in the volume of the sensing layer. During the interaction H- bonds are formed between hydroxyl groups of target molecules and carboxylic groups in polymer matrix and these interactions highly depend on the functional groups present in the target molecules. The number of functional groups in TC and OTC molecules is different and this difference is responsible for the difference in the sensitivity.

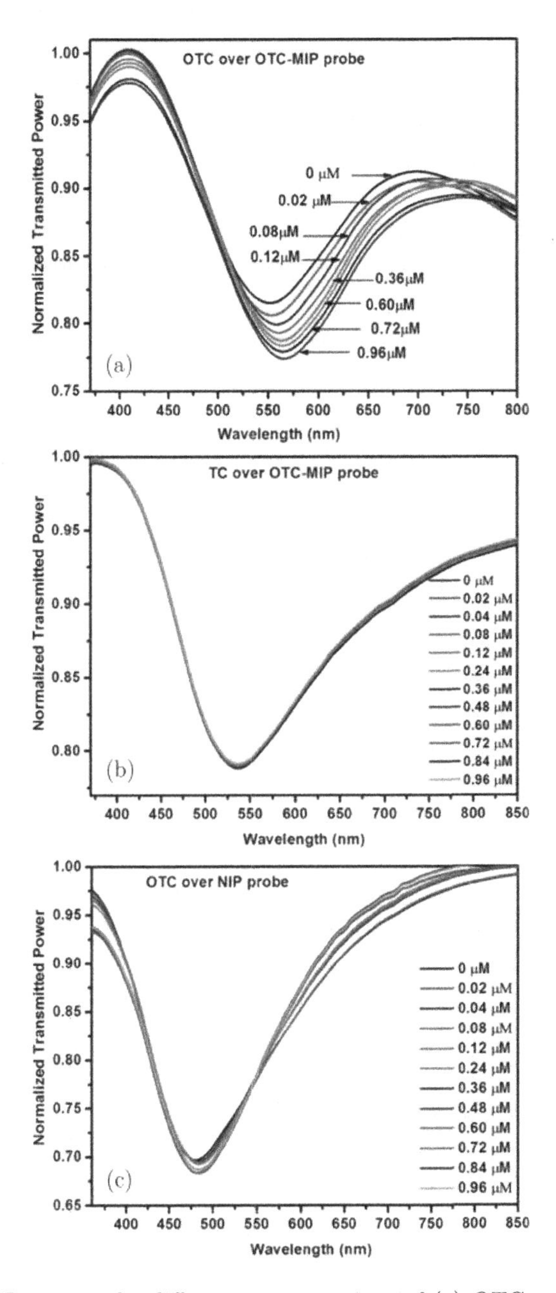

Fig. 6.31. SPR spectra for different concentrations of (a) OTC over OTC–MIP probe, (b) TC over OTC–MIP probe and (c) OTC over NIP probe. Reprinted from Ref. [19] with permission from Royal Society of Chemistry.

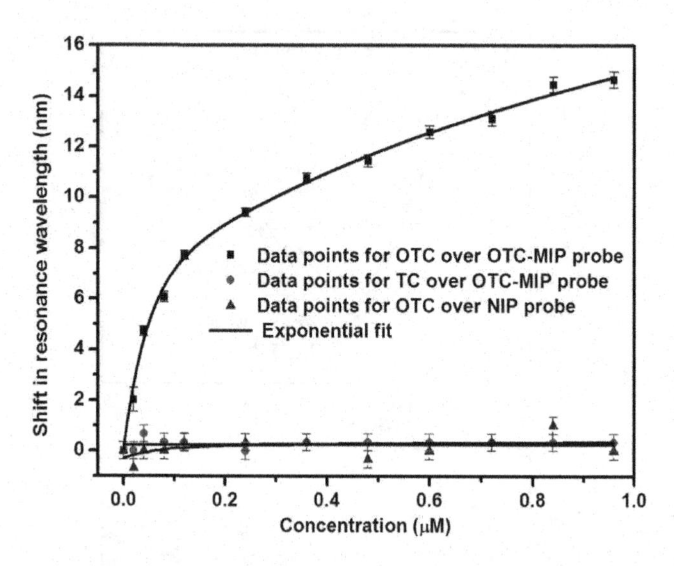

Fig. 6.32. Variation of shift in resonance wavelength with the concentration of OTC and TC samples. Reprinted from Ref. [19] with permission from Royal Society of Chemistry.

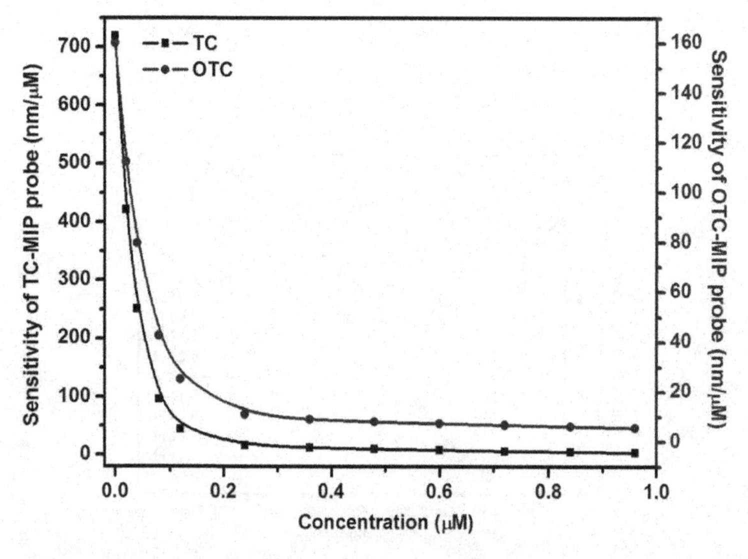

Fig. 6.33. Variation of sensitivity of the TC–MIP probe and OTC–MIP probe with the concentration of TC and OTC respectively. Reprinted from Ref. [19] with permission from Royal Society of Chemistry.

6.7 Multi-analyte Sensing

In Sec. 6.2.2 we have discussed the detection of refractive index at multiple places or of multiple samples using SPR based fiber optic probe. The sensor is named as multi-channel fiber optic sensor. In this section, we shall discuss a fiber optic multi-analyte sensor using SPR technique. The multi-analyte means that the sensor can detect more than one analyte present in a fluid sample using a single set up. Recently a fiber optic multi-analyte sensor utilizing SPR technique has been reported for simultaneous detection of glucose and urea in a fluid sample.[20] In Secs. 6.5.1 and 6.5.5 we have discussed independently the fiber optic urea and glucose sensors respectively. Here we discuss fiber optic SPR sensor which can detect both urea and glucose simultaneously in a sample. The procedure of probe fabrication is the same as discussed before. The cladding was removed from two different locations of an optical fiber. One unclad portion was used to fabricate urea sensor, while the other unclad region was for the fabrication of the glucose sensor. Two unclad core were coated with different metals to obtain two well-separated dips in the SPR spectrum. The scanning electron microscope images of the surfaces with and without immobilized enzymes were taken to see the difference between the coatings on the surface. These images for no enzyme immobilized and enzyme immobilized sensing surfaces are shown in Fig. 6.34. Figures 6.34(a), (b), and (c) show no enzyme immobilized surface, urease immobilized surface and glucose oxidase (GOD) immobilized surface, respectively.[20] The no enzyme immobilized surface is a plane surface while there are some pores of different sizes on the enzyme immobilized sensing surfaces.

The experimental set up of the two channel SPR based fiber optic sensor for the detection of urea and glucose (channel 1 and channel 2) simultaneously is shown in Fig. 6.35. The fiber optic sensing probe fabricated with two channels is fixed in a single glass flow cell for the detection of two analytes present in one fluid sample. The SPR spectra recorded for mixtures of urea and glucose in different proportions are shown in Fig. 6.36. Two well separated SPR dips, one corresponding to channel 1 and another corresponding to channel 2

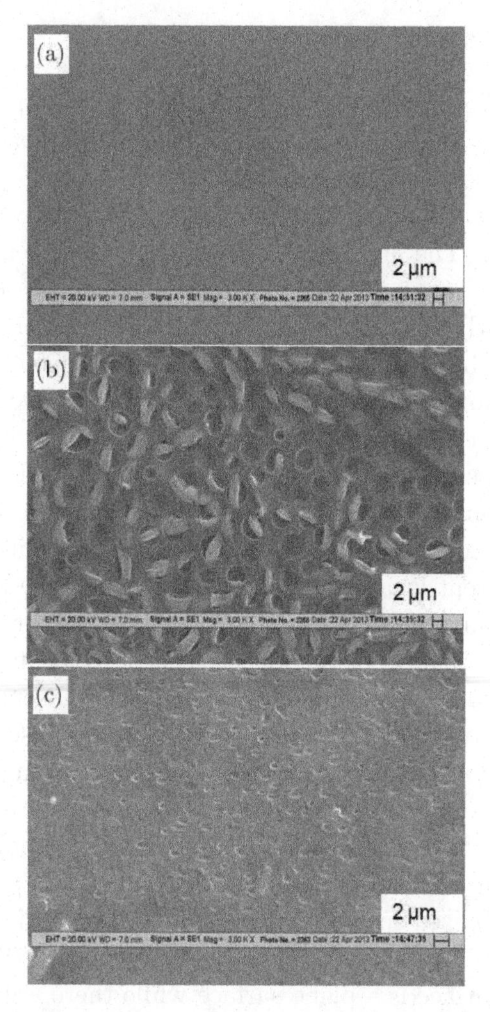

Fig. 6.34. Scanning Electron Microscope images of surfaces with (a) no enzyme entrapped gel (b) urease entrapped gel, and (c) glucose oxidase entrapped gel. Reprinted from Ref. [20] with permission from Royal Society of Chemistry.

can be seen. To distinguish the analytes of the dips both the channels were exposed to the sample having only one of the two analytes.

Figure 6.37(a) shows the SPR spectra for fluid samples of different concentrations of urea only. The SPR dips at lower wavelength side (or left side) shift towards blue side while no shift in SPR dips at higher wavelength side (or right side) was observed with

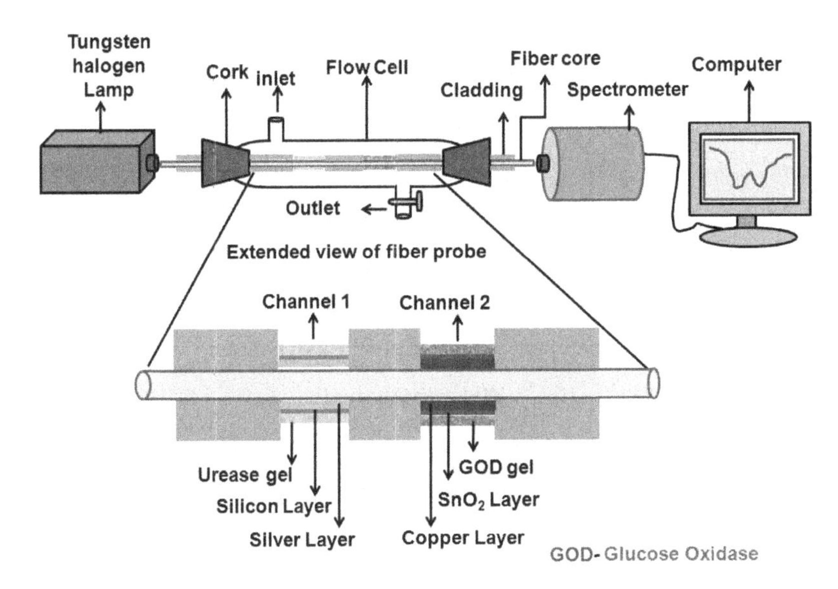

Fig. 6.35. Experimental set up of SPR based fiber optic two channels, urea and glucose, sensor. Reprinted from Ref. [20] with permission from Royal Society of Chemistry.

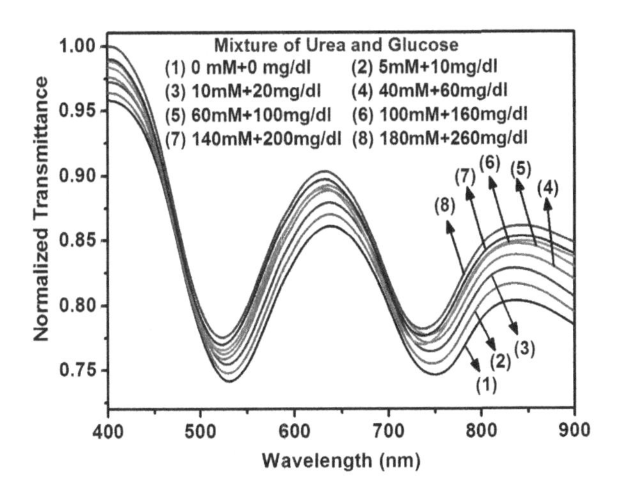

Fig. 6.36. SPR spectra of fiber optic dual channel sensing probe for different concentration mixtures of urea and glucose in buffer in the range 0–180 mM and 0–260 mg/dl respectively. Reprinted from Ref. [20] with permission from Royal Society of Chemistry.

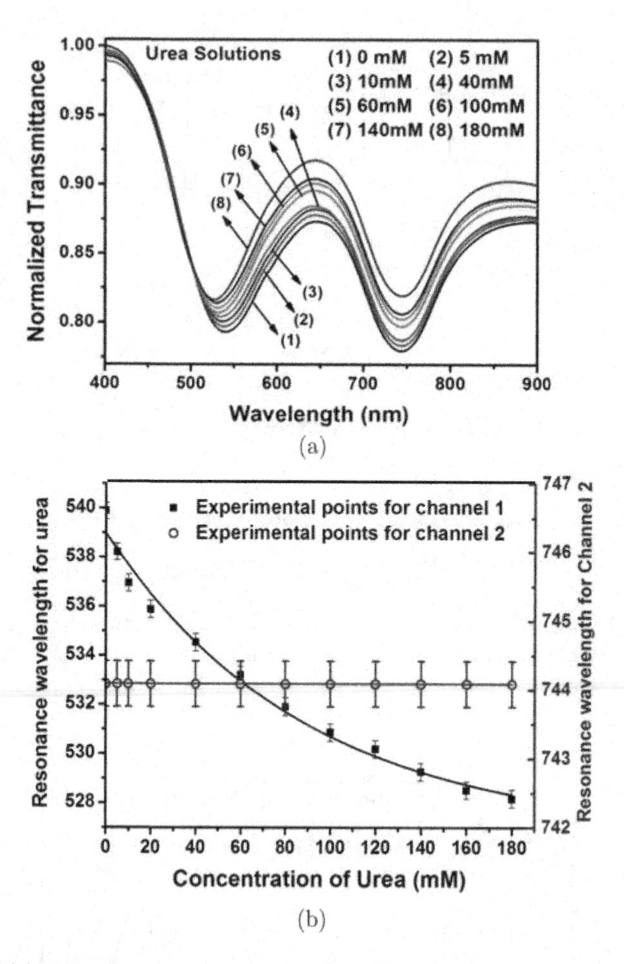

Fig. 6.37. (a) SPR spectra of fiber optic dual channel sensing probe for the different concentrations of urea in buffer solutions with no glucose added. (b) Variation of resonance wavelengths of channel 1 and channel 2 determined from the SPR spectra with the concentration of urea in buffer. Reprinted from Ref. [20] with permission from Royal Society of Chemistry.

the increasing concentration of urea. This implies that the SPR dips at lower wavelength side are due to urea and correspond to channel 1. The SPR dips at higher wavelength side where no shift is observed correspond to glucose sensing and channel 2. The variation of resonance wavelengths for channel 1 and channel 2 with the concentration of urea are shown in Fig. 6.37(b). The resonance wavelength in channel 1 decreases with the increasing concentration of urea which is due to the reaction of urea with urease resulting in the decrease in the effective refractive index of the sensing gel layer of channel 1. For channel 2, the resonance wavelength is nearly constant because no reaction takes place between enzyme GOD present in gel layer and urea present in the sample because enzymes are very selective to analytes.

The SPR spectra for different concentrations of glucose samples around both the channels are shown in Fig. 6.38(a). Similar to channel 1 in Fig. 6.37(a), in this case, blue shift of SPR dips of channel 2 occurs on increasing the concentration of glucose while no shift of the SPR dips of channel 1 appears to occur. This implies that channel 1 corresponds to detection of urea and channel 2 corresponds to the detection of glucose. The resonance wavelengths determined from SPR spectra of Fig. 6.38(a) for both the channels for different concentrations of glucose are shown in Fig. 6.38(b). The decrease in resonance wavelength with the increasing concentration of glucose for channel 2 occurs because when glucose comes in contact with enzyme (GOD) immobilized gel layer a decrease in the effective refractive index of sensing layer occurs due to the conversion of glucose to gluconic acid and H_2O_2 in the presence of GOD. Since no reaction takes place between glucose and enzyme urease immobilized gel layer in channel 1 no shift in the resonance wavelength is observed for channel 1. The sensing activities of both the channels appear to be independent of each other. This is mainly because enzymes are very selective to corresponding analytes.

Let us go back to SPR spectra of Fig. 6.36 where the sample contains both urea and glucose. The resonance dips at lower wavelength correspond to urea detection while the SPR dips at higher wavelength correspond to glucose detection. The variations of resonance wavelengths of channel 1 and channel 2 with the concentration of urea and glucose are shown in Figs. 6.39(a) and 6.39(b) respectively. Both the variations are non-linear and resonance wavelength of both the channels decreases asymptotically with the increasing concentration of corresponding analyte and

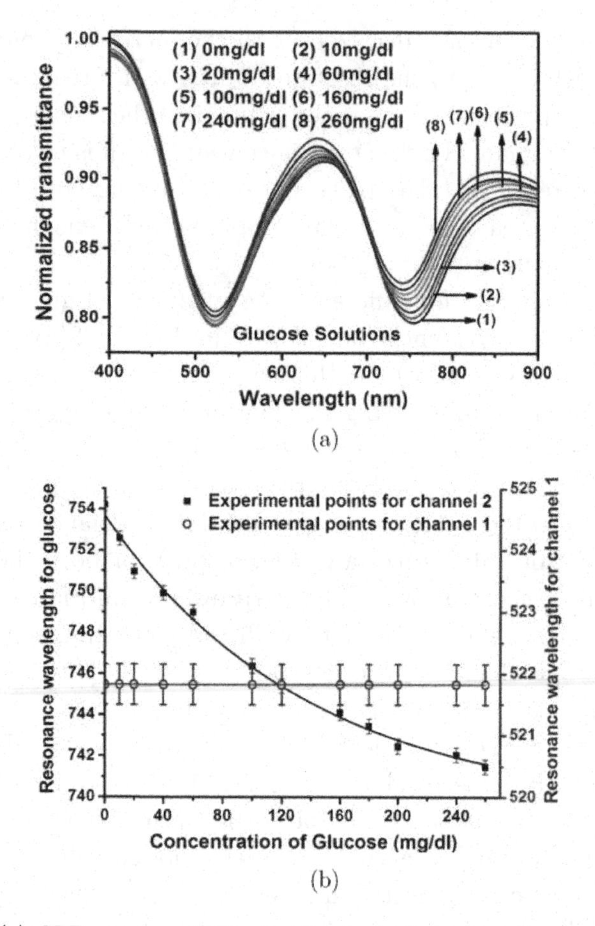

Fig. 6.38. (a) SPR spectra of fiber optic dual channel sensing probe for the different concentrations of glucose in buffer solutions with no urea added. (b) Variation of resonance wavelength of channel 1 and channel 2 determined from the SPR spectra with concentration of glucose in buffer. Reprinted from Ref. [20] with permission from Royal Society of Chemistry.

appears to get saturated at the higher concentrations of the analyte. The saturation of resonance wavelength at higher concentration occurs due to the limited quantities of enzymes (urease and GOD) in the gels in respective channels as discussed earlier in this chapter.

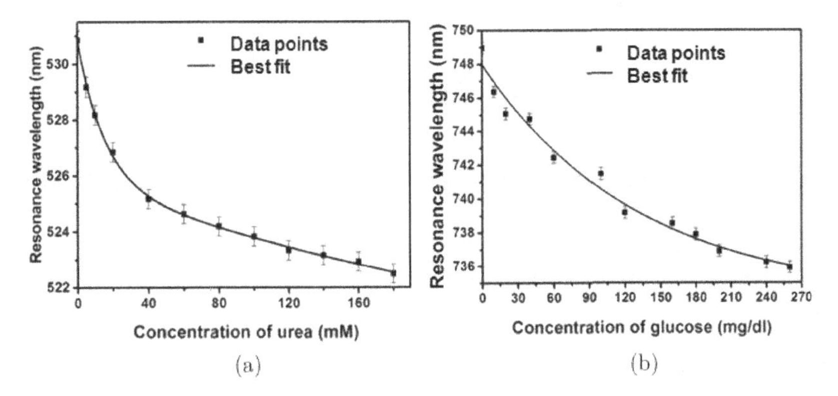

Fig. 6.39. Variation of resonance wavelength with the concentration of (a) urea and (b) glucose in the mixture prepared in buffer for dual channel sensing probe. The resonance wavelengths are determined from the SPR spectra of Fig. 6.36. Reprinted from Ref. [20] with permission from Royal Society of Chemistry.

6.8 Gas Sensors

So far we have discussed detection of analytes in the form of aqueous solutions. However, SPR technique along with optical fiber can also be used for the detection of various gases. Since the refractive index of the gases is low, these cannot be sensed using conventional SPR sensor. Coating a layer of some material which changes its optical properties in the presence of the gas is required over metal layer in the conventional SPR sensor. Metal oxides such as ZnO, tin oxide (SnO_2), and indium oxide (In_2O_3) have been used for the gas sensing.[21–23] ITO thin film exhibits better performance than gold or silver film if used as plasmonic metal and sensing of gases.[24,25] In the case of ITO, there are no band-to-band transitions and islands formation even on a very thin layer of ITO deposited on the dielectric surface, the problem faced with gold or silver layer. ITO has been used in gas sensors to detect gases such as carbon monoxide (CO), H_2, nitrogen dioxide (NO_2), etc. The specific disadvantage of ITO coated SPR sensor is its operability at high temperature and some restrictions in sensitivity, selectivity, and response time. Conducting polymers such as polyaniline (PANi) have also been used as active layer for gas sensing.[26] These polymers have high sensitivity and short response time even at room temperature. These can be easily synthesized and have good mechanical properties which allow an easy fabrication of sensors. Attention has also been paid to the sensors fabrication using conducting polymers. In this section we shall discuss the detection of ammonia (NH_3) and hydrogen sulphide (H_2S) gases using SPR technique.

6.8.1 *Ammonia gas sensor*

Ammonia or NH_3 gas is a colorless gas with pungent odor. Main sources of NH_3 are the combustion in motor vehicles and chemical plants. Moreover it is extensively used in explosives, fertilizers, and as an industrial coolant. Its inhalation may cause poisoning to people. Several optical approaches have been used for the fabrication of NH_3 sensor but optical fiber based SPR sensors present a cheap and viable option for the detection of NH_3 gas because of the various advantages mentioned in previous chapters. Fabrication and characterization of a SPR based fiber optic NH_3 sensor using bromocresol purple (BCP) as transducing layer were reported.[27] The probe was fabricated by depositing layers of silver, silicon, and BCP over an unclad portion of the fiber. As discussed earlier, silicon has been used to protect metallic layer from oxidation. The experimental set up of a SPR based fiber optic gas sensor is shown in Fig. 6.40.

The sensing is based on the change in the refractive index of the BCP dye layer due to the change in NH_3 gas concentration around the probe resulting in the shift in resonance wavelength. In addition to change in refractive index, the change in the color of the dye also occurs which is probably due to the florescence properties of the dye. The SPR spectra obtained for this probe are shown in Fig. 6.41. The SPR spectra show peaks as well as dips. The peak corresponds to fluorescence while the dip is due to SPR phenomenon. The location of dip shifts towards the higher wavelength as the concentration of NH_3 gas in the chamber increases which can be seen more precisely in Fig. 6.42 having plot of the resonance wavelength

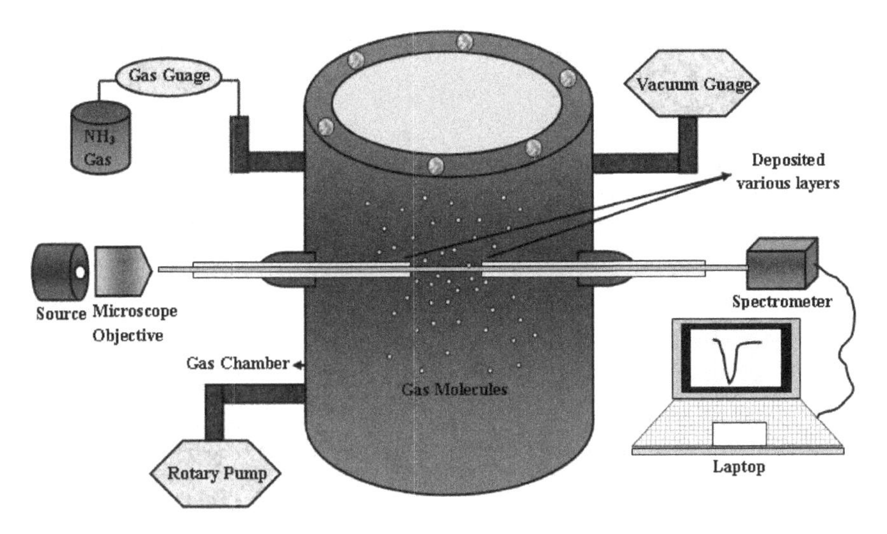

Fig. 6.40. Experimental set up of a SPR based fiber optic NH_3 gas sensor. Reprinted from Ref. [27] with permission from Springer.

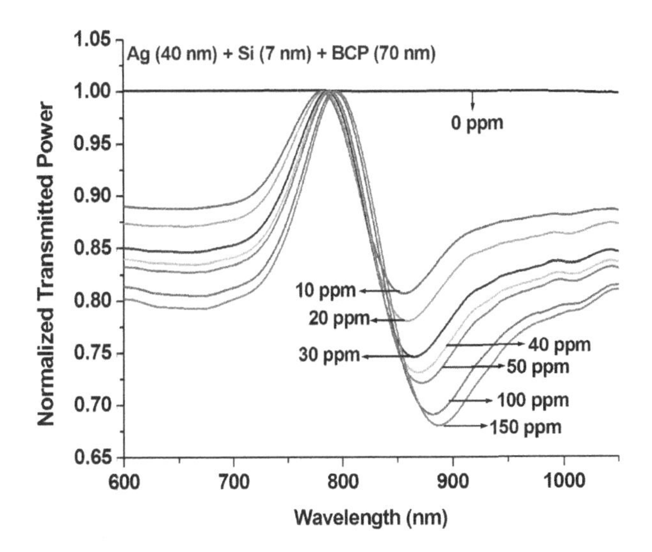

Fig. 6.41. SPR spectra of fiber optic probe having layers of silver, silicon, and BCP over unclad core of the fiber for different concentrations of NH_3 gas. Reprinted from Ref. [27] with permission from Springer.

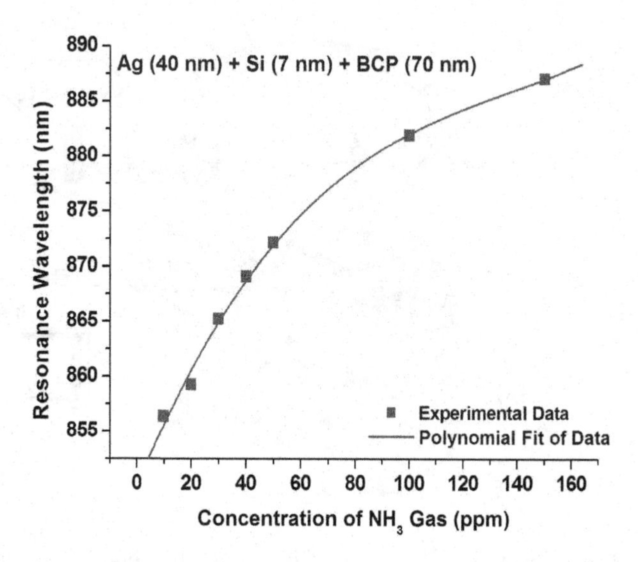

Fig. 6.42. Variation of resonance wavelength with the concentration of NH_3 gas. Reprinted from Ref. [27] with permission from Springer.

with the concentration of NH_3 gas. The increase in the thickness of the silicon layer or the BCP layer shifts the SPR spectra towards higher wavelength implying that the operating range of the sensor can be tuned by varying the thicknesses of the silicon/BCP layer.

In another SPR based fiber optic NH_3 gas sensor unclad core is coated with the layers of ITO and PANi.[28] PANi, a conducting polymer, changes its refractive index on exposure to NH_3 gas and is a highly environmentally stable material. For the fabrication of the fiber optic SPR probe, a small portion of the unclad core of the fiber was coated with a thin film of ITO. Further, this layer was dip coated in PANi solution which was chemically prepared by polymerization of aniline monomer. When NH_3 gas comes in contact of the PANi film, it changes the optical properties of the film and hence the shift in SPR dip is recorded. The thickness of the ITO film was optimized for the best performance of the sensor. It was shown that the SPR sensor with PANi coated over ITO film exhibits higher sensitivity than PANi coated over gold film.

Recently SPR based fiber optic NH_3 gas sensor utilizing nano-composite film of poly(methyl methacrylate) [PMMA] and reduced graphene oxide (rGO) over copper film coated unclad core of the optical fiber was also reported.[29] The SPR spectra of the probe shift towards red side on increase in the concentration of NH_3 gas in the chamber. The sensitivity of the sensor depends on the doping concentration of reduced graphene in the nanocomposite.

6.8.2 *Hydrogen sulphide gas sensor*

Hydrogen sulphide or H_2S is a colorless, explosive and highly toxic gas. It can affect human nervous system and could cause the people to lose consciousness. H_2S is released from sewage sludge, liquid manure, and sulphur hot springs. It is a by-product of the many industrial processes including petroleum refining and is produced naturally by the decay of organic matter. As far as its applications are concerned it is used to produce elemental sulphur, sulphuric acid, and heavy water for the nuclear reactors. A SPR based fiber optic sensor for the detection of H_2S gas using coatings of copper and zinc oxide (ZnO) thin films over unclad core was reported.[30] The properties such as dielectric function of ZnO are highly sensitive towards H_2S. As the H_2S gas comes in contact with the ZnO layer in the probe, it changes the dielectric constant or the refractive index of the ZnO layer which in turn changes the resonance wavelength of the SPR spectrum. The fabricated probe is fixed in a gas chamber and the experimental set up of the sensor is similar to that shown in Fig. 6.40.

Figure 6.43 shows the SPR spectra of the fiber optic SPR probe for the concentration of H_2S gas varying from 10 ppm to 100 ppm at room temperature. As the concentration of H_2S gas increases the SPR curve shifts towards higher wavelength. In other words, increase in the concentration of the H_2S gas increases the resonance wavelength. This is because of the increase in the refractive index of the ZnO layer on exposure to H_2S gas. When H_2S gas comes in contact with the ZnO following reaction takes place

$$ZnO_{(s)} + H_2S_{(g)} \rightleftharpoons ZnS_{(s)} + H_2O_{(g)}$$

The ZnS formed in the reaction covers the surface of the ZnO grains which in turn results in the change in the refractive index of the ZnO film coated over sensing probe. In addition to shift

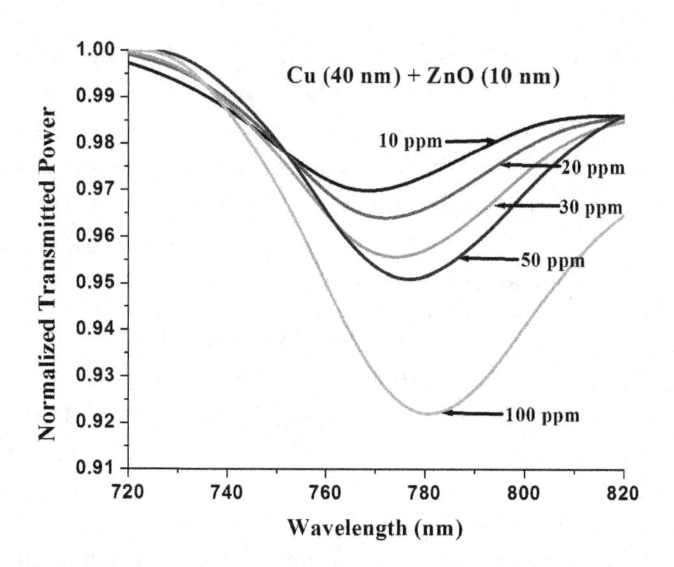

Fig. 6.43. SPR spectra of fiber optic SPR probe for different concentrations of the H_2S gas. The probe has layers of 40 nm thin copper and 10 nm thin ZnO. Reprinted from Ref. [30] with permission from Royal Society of Chemistry.

of SPR curve, the depth of the SPR curve also increases with the increase in the concentration of H_2S gas. This implies that both the real and imaginary parts of the refractive index of the ZnO changes with the change in the concentration of the H_2S gas. The change in the real part changes the resonance wavelength whereas the change in the imaginary part changes the absorbance and hence the depth of the SPR curve. The variation of the resonance wavelength determined from the SPR spectra of Fig. 6.43 with the concentration of the H_2S gas is shown in Fig. 6.44. The resonance wavelength increases with the increase in the concentration of the H_2S gas and appears to saturate beyond 100 ppm concentration. Initially the rate of increase of resonance wavelength is high, but decreases as the concentration further increases. This is because at low concentration of the gas, the available ZnO molecules per H_2S molecule is large so the change in the resonance wavelength is more but as the concentration of the H_2S increases the number of ZnO molecules per H_2S molecule decreases. Hence, at higher concentration of the H_2S gas the resonance wavelength saturates. In terms of optical

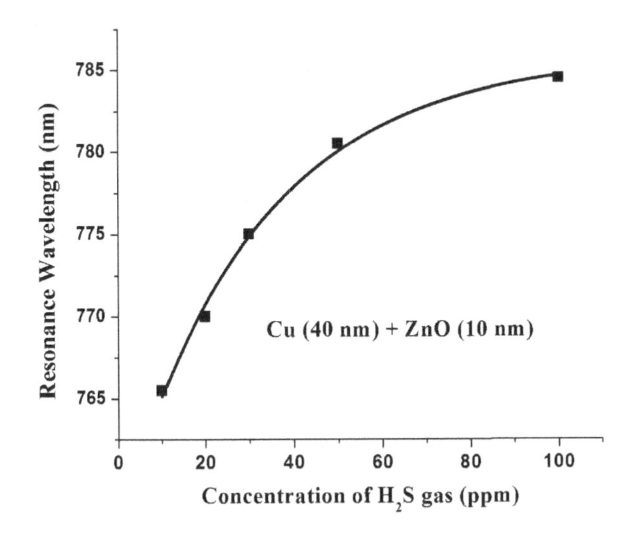

Fig. 6.44. Variation of resonance wavelength with the concentration of H_2S gas for the SPR probe having layers of copper (40 nm), and ZnO (10 nm). Reprinted from Ref. [30] with permission from Royal Society of Chemistry.

properties, the refractive index of ZnO increases with the increase in the concentration of H_2S gas and becomes constant after nearly 100 ppm concentration of the gas.

The variation of sensitivity of the sensor, defined as the shift in resonance wavelength per unit change in the concentration of the gas, as a function of the concentration of the gas is shown in Fig. 6.45. The sensitivity decreases with the increase in the concentration of the H_2S gas and becomes negligible at higher concentrations of the H_2S gas (greater than 100 ppm). The physical reason behind this is the same as mentioned for Fig. 6.44.

Various other metal oxides were also tested for the sensing of H_2S gas. These are TiO_2, SiO_2, and SnO_2. The total shift in resonance wavelength for the concentration range 10 ppm to 100 ppm for the probes with the layers of these oxides are given in Fig. 6.46. The total shift in resonance wavelength for all the fabricated probes are determined from their SPR spectra for these two concentrations. The histogram shows that the maximum shift is for the probe having ZnO coating over the copper layer. The probe with TiO_2 coating has minimum amount of shift in resonance wavelength.

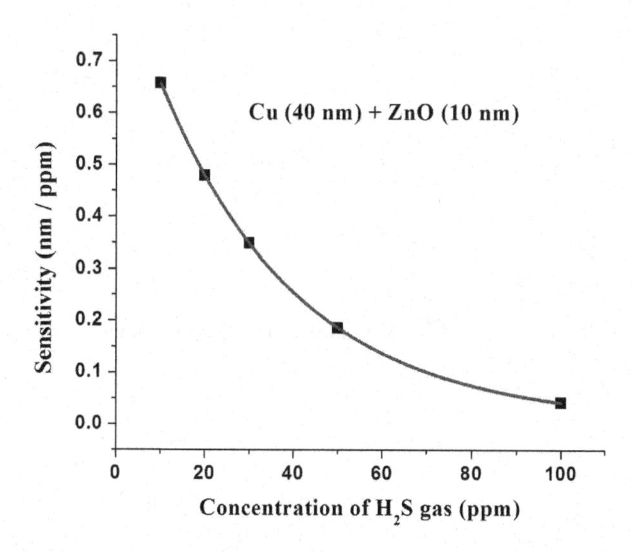

Fig. 6.45. Variation of sensitivity of the sensor with the concentration of the H_2S gas for the SPR probe having layers of copper (40 nm), and ZnO (10 nm). Reprinted from Ref. [30] with permission from Royal Society of Chemistry.

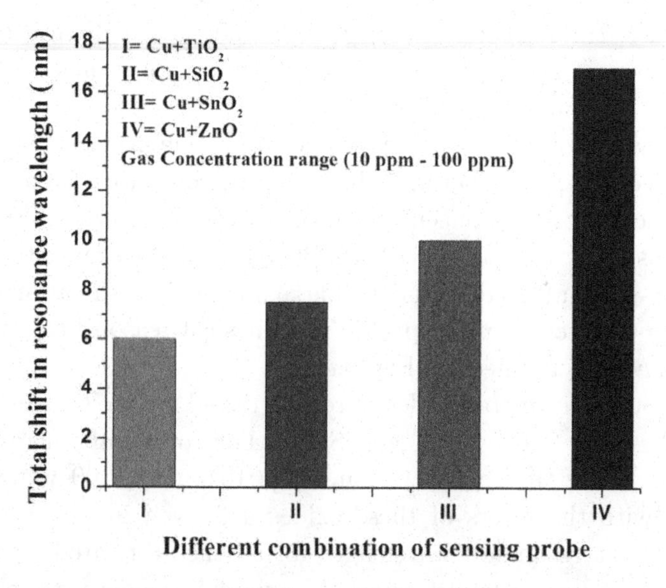

Fig. 6.46. Shift in resonance wavelength for the concentration range of H_2S gas from 10 ppm to 100 ppm for SPR probes having coatings of different metal oxides over copper layer. Reprinted from Ref. [30] with permission from Royal Society of Chemistry.

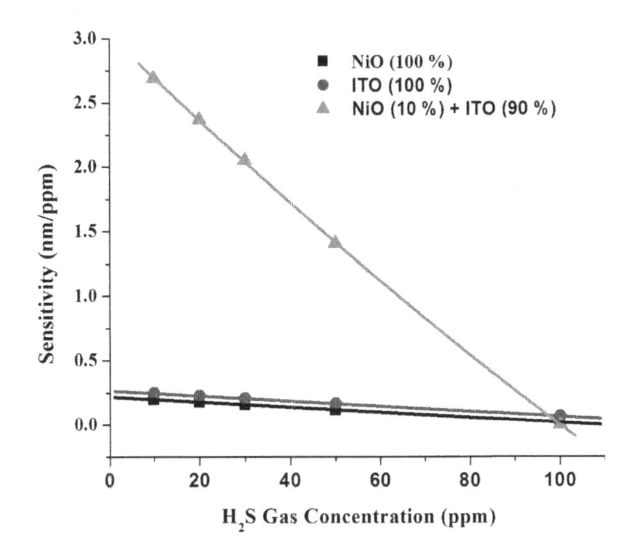

Fig. 6.47. Variation of sensitivity with the concentration of the H_2S gas in the range 10–100 ppm for the SPR probes having layers of ITO (7 nm), NiO (7 nm) and NiO doped ITO (7 nm). Reprinted from Ref. [31] with permission from Elsevier.

Another SPR based fiber optic H_2S gas sensor reported uses nickel oxide (NiO) doped ITO thin film over the silver coated unclad core of the fiber.[31] The experimental set up of the sensor is similar to that shown in Fig. 6.40. The variation of sensitivity with the concentration of H_2S in the range 10–100 ppm for the NiO doped ITO coated probe is shown in Fig. 6.47. In the same figure the variation has also been shown for the probes having over-layers of NiO and ITO over silver coated unclad core of the fiber. The sensitivity is not a constant. It decreases with the increase in the concentration of the gas for all the three probes. However, it is maximum for the NiO doped ITO film.

The int eraction of other gases with NiO doped ITO coated probe for the high concentration of the gases was also studied. Figure 6.48(a) shows the total shift in resonance wavelength for the change in concentration of the gases from 10 to 100 ppm. The figure shows that the NiO doped ITO film interacts with all the gases although the shift in the resonance wavelength is different for different gases. The total shift in the resonance wavelength in the

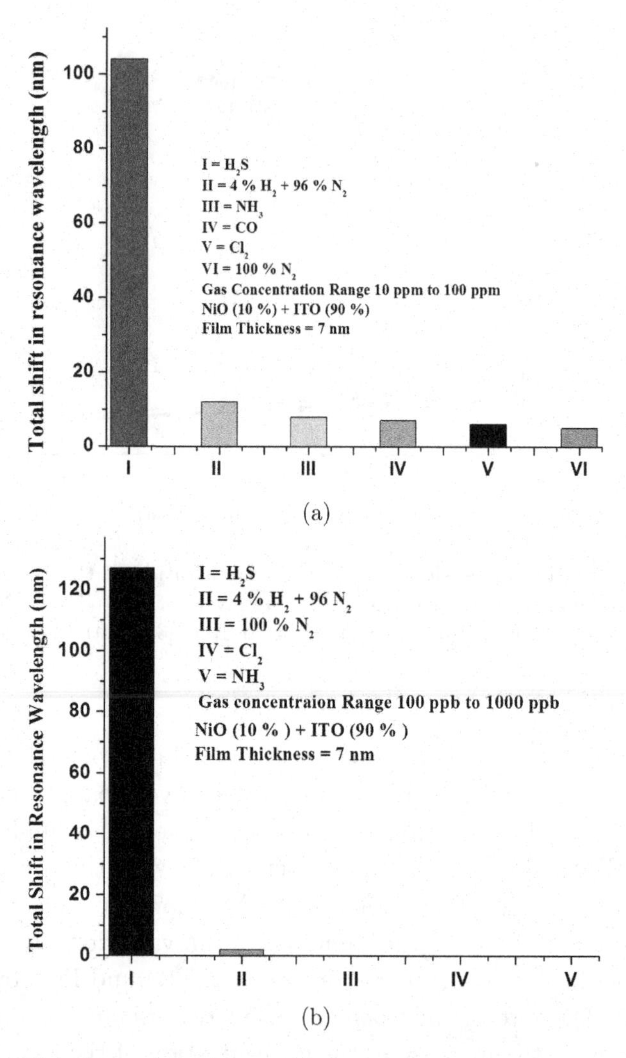

Fig. 6.48. Total shift in the resonance wavelength for the concentration range (a) 10–100 ppm and (b) 100–1000 ppb in the case of different gases around NiO doped ITO over-layer as a sensing layer. Reprinted from Ref. [31] with permission from Elsevier.

case of NiO doped ITO coated probe is maximum for the H_2S gas as compared to the other gases. The interaction of these gases with the NiO doped ITO coated probe for the low concentrations of the gases was also studied. Figure 6.48(b) shows the total shift in the resonance wavelength for the concentration range 100–1000 ppb of the different gases. The histogram shows that the NiO doped ITO thin film coated probe is highly selective for H_2S gas for the low concentration range of the H_2S gas.

6.9 Summary

In this chapter we have discussed various SPR based fiber optic sensors for the detection of different chemical and biological analytes such as pH, ethanol, urea, glucose, naringin, pesticide, phenolic compounds, low density lipoprotein, vitamin B_3, TC, and gases like NH_3 and H_2S. The sensors of these analytes are either based on conventional SPR sensor with entrapment of enzyme in gel layer coated over metal layer or are based on MIP where binding sites are prepared by target itself for the analyte to be sensed. The enhancement of sensitivity of the refractive index sensor using high refractive index layer has been discussed. In addition multi-channel and multi-analyte sensors have also been presented.

References

1. P. Bhatia and B.D. Gupta, Surface-plasmon-resonance-based fiber-optic refractive index sensor; sensitivity enhancement, *Applied Optics* **50** (2011) 2032–2036.
2. R.C. Jorgenson and S.S. Yee, A fiber-optic chemical sensor based on surface plasmon resonance, *Sensors and Actuators B* **12** (1993) 213–220.
3. A. Lahav, M. Auslender and I. Abdulhalim, Sensitivity enhancement of guided-wave surface plasmon resonance sensors, *Optics Letters* **33** (2008) 2539–2541.
4. S. Singh, S.K. Mishra and B.D. Gupta, Sensitivity enhancement of a surface plasmon resonance based fibre optic refractive index sensor utilizing an additional layer of oxides, *Sensors and Actuators A* **193** (2013) 136–140.
5. R. Verma and B.D. Gupta, SPR based three channels fiber optic sensor for aqueous environment, *Proceedings of SPIE* 8992 (2014) 89920A.
6. B.D. Gupta and N.K. Sharma, Fabrication and characterization of U-shaped fiber-optic pH probes, *Sensors and Actuators B* **82** (2002) 89–93.
7. J.M. Corres, I.R. Matias, I.D. Villar and F.J. Arregui, Design of pH sensors in long period fiber gratings using polymeric nanocoatings, *IEEE Sensors Journal* **7** (2007) 455–463.

8. I. Yulianti, A.S.M. Supa'at, S.M. Idrus, O. Kurdi and M.R.S. Anwar, Sensitivity improvement of a fiber Bragg grating pH sensor with elastomeric coating, *Measurement Science and Technology* **23** (2012) 015104.
9. S. Singh and B.D. Gupta, Fabrication and characterization of a highly sensitive surface plasmon resonance based fiber optic pH sensor utilizing high index layer and smart hydrogel, *Sensors and Actuators B* **173** (2012) 268–273.
10. S.K. Mishra and B.D. Gupta, Surface plasmon resonance based fiber optic pH sensor utilizing Ag/ITO/Al/hydrogel layers, *Analyst* **138** (2013) 2640–2646.
11. S.K. Srivastava, R. Verma and B.D. Gupta, Surface plasmon resonance based fiber optic sensor for the detection of low water content in ethanol, *Sensors and Actuators B* **153** (2011) 194–198.
12. P. Bhatia and B.D. Gupta, Fabrication and characterization of a surface plasmon resonance based fiber optic urea sensor for biomedical applications, *Sensors and Actuators B* **161** (2011) 434–438.
13. Rajan, S. Chand and B.D. Gupta, Fabrication and characterization of a surface plasmon resonance based fiber-optic sensor for bittering component — Naringin, *Sensors and Actuators B* **115** (2006) 344–348.
14. Rajan, S. Chand and B.D. Gupta, Surface plasmon resonance based fiber-optic sensor for the detection of pesticide, *Sensors and Actuators B* **123** (2007) 661–666.
15. S. Singh, S.K. Mishra and B.D. Gupta, SPR based fibre optic biosensor for phenolic compounds using immobilization of tyrosinase in polyacrylamide gel, *Sensors and Actuators B* **186** (2013) 388–395.
16. S. Singh and B.D. Gupta, Fabrication and characterization of a surface plasmon resonance based fiber optic sensor using gel entrapment technique for detection of low glucose concentration, *Sensors and Actuators B* **177** (2013) 589–595.
17. R. Verma, S.K. Srivastava and B.D. Gupta, Surface-plasmon-resonance-based fiber-optic sensor for the detection of low-density lipoprotein, *IEEE Sensors Journal* **12** (2012) 3460–3466.
18. R. Verma and B.D. Gupta, Fiber optic SPR sensor for the detection of 3-pyridinecarboxamide (vitamin B_3) using molecularly imprinted hydrogel, *Sensors and Actuators B* **177** (2013) 279–285.
19. R. Verma and B.D. Gupta, Optical fiber sensor for the detection of tetracycline using surface plasmon resonance and molecular imprinting, *Analyst* **138** (2013) 7254–7263.
20. R. Verma and B.D. Gupta, A novel approach for simultaneous sensing of urea and glucose by SPR based optical fiber multi-analyte sensor, *Analyst* **139** (2014) 1449–1455.
21. S.K. Gupta, A. Joshi and M. Kaur, Development of gas sensors using ZnO nanostructures, *Journal of Chemical Science* **122** (2010) 57–62.
22. G.E. Patil, D.D. Kajale, D.N. Chavan, N.K. Pawar, P.T. Ahire, S.D. Shinde, V.B. Gaikwad and G.H. Jain, Synthesis, characterization and gas sensing

performance of SnO_2 thin films prepared by spray pyrolysis, *Bulletin of Material Science* **34** (2011) 1–9.
23. M. Suchea, N. Katsarakis, S. Christoulakis, S. Nikolopoulo and G. Kiriakidis, Low temperature indium oxide gas sensors, *Sensors and Actuators B* **118** (2006) 135–141.
24. C. Rhodes, M. Cerruti, A. Efremenko, M. Losego, D.E. Aspnes, J.P. Maria and S. Franzen, Dependence of plasmonpolaritons on the thickness of ITO thin films, *Journal of Applied Physics* **103** (2008) 093108.
25. S. Franzen, C. Rhodes, M. Cerruti, R.W. Gerber, M. Losego, J.P. Maria and D.E. Aspens, Plasmonic phenomenon in ITO and ITO–Au hybrid films, *Optics Letters* **34** (2009) 2867–2869.
26. D. Nicolas-Debarnot and F. Poncin-Epaillard, Polyaniline as a new sensitive layer for gas sensors, *Analytica Chimica Acta* **475** (2003) 1–15.
27. P. Bhatia and B.D. Gupta, Surface plasmon resonance based fiber optic ammonia sensor utilizing bromocresol purple, *Plasmonics* **8** (2013) 779–784.
28. S.K. Mishra, D. Kumari and B.D. Gupta, Surface plasmon resonance based fiber optic ammonia gas sensor using ITO and polyaniline, *Sensors and Actuators B* **171–172** (2012) 976–983.
29. S.K. Mishra, S.N. Tripathi, V. Choudhary and B.D. Gupta, SPR based fibre optic ammonia gas sensor utilizing nanocomposite film of PMMA/reduced graphene oxide prepared by in situ polymerization, *Sensors and Actuators B* **199** (2014) 190–200.
30. R. Tabassum, S.K. Mishra and B.D. Gupta, Surface plasmon resonance-based fiber optic hydrogen sulphide gas sensor utilizing Cu–ZnO thin films, *Physical Chemistry Chemical Physics* **15** (2013) 11868–11874.
31. S.K. Mishra, S. Rani and B.D. Gupta, Surface plasmon resonance based fiber optic hydrogen sulphide gas sensor utilizing nickel oxide doped ITO thin film, *Sensors and Actuators B* **195** (2014) 215–222.

Chapter 7

SPR based Fiber Optic Sensors: Factors Affecting Performance

7.1 Introduction

The performance of surface plasmon resonance (SPR) based fiber optic sensors depends on various factors. These include the choice of optical fiber, metals for coating, dopants in the core of the fiber, addition of a high index dielectric layer between metal and sensing medium, change in the probe design, and the change in the external stimuli such as temperature and the polarity of the sensing sample. In this chapter we shall discuss the effects of these parameters on various performance parameters of the SPR based fiber optic sensors.

7.2 Influence of Intrinsic Stimuli

The intrinsic stimuli are the fiber parameters such as core diameter, sensing length and numerical aperture, metal layer, dopants in the fiber core, probe modification using additional layers, and probe geometry such as tapered and U-shaped. We shall discuss the effects of these stimuli on the performance parameters such as calibration curve, sensitivity, and figure of merit of the sensor.

7.2.1 *Fiber parameters*

The crucial fiber parameters are core diameter, sensing length, and the numerical aperture. The effects of these parameters on sensor performance have been reported in Ref. [1] in detail. We shall discuss their effects one by one.

7.2.1.1 *Core diameter*

The role of core diameter is in the number of reflections a ray makes in the sensing region which is given by Eq. (4.34). To see its effect we consider collimated source and focusing lens combination for launching of light in the fiber. In other words, we assume propagation of only all guided meridional rays in the fiber. Further, we use the following values of the fiber parameters and gold metal layer for simulations: numerical aperture (NA) of the fiber = 0.25, $L = 1\,\text{cm}$, metal layer thickness = 45 nm, and $n_s = 1.330$. The diameter of the fiber core is varied from 400 μm to 600 μm. Figures 7.1 to 7.3 show the variation of sensitivity, spectral width of SPR spectrum, and the figure of merit as a function of fiber core diameter, respectively. The figures show that as the core diameter increases the sensitivity and figure of merit of the sensor increase while the width of the SPR curve decreases. This is because the number of reflections of a ray in the fiber core decreases as the core diameter increases.

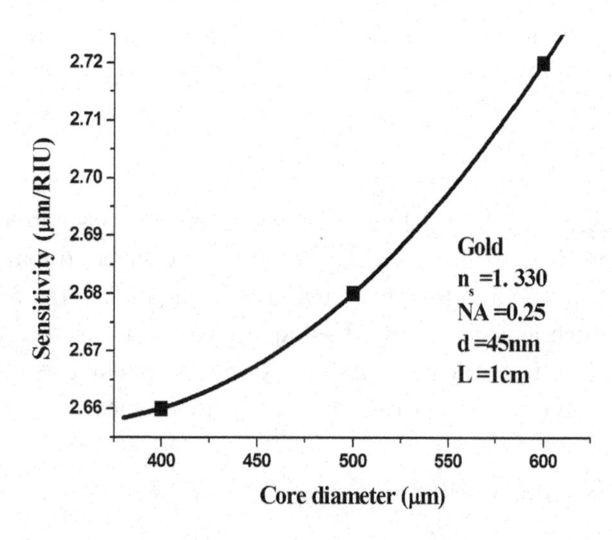

Fig. 7.1. Variation of sensitivity of the SPR based fiber optic sensor with the diameter of the fiber core.

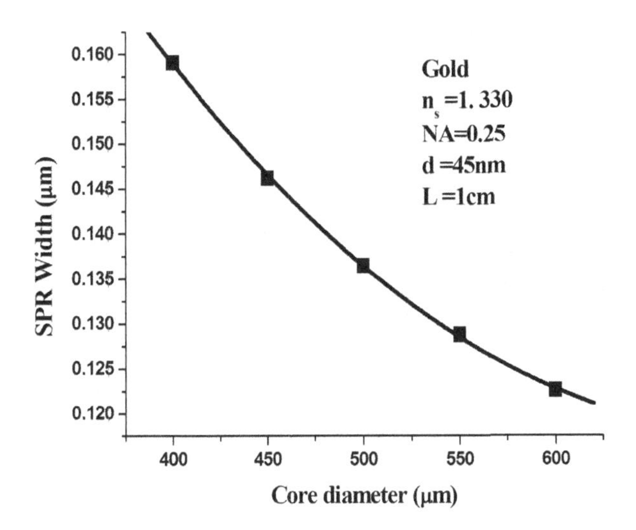

Fig. 7.2. Variation of spectral width of the SPR curve of the SPR based fiber optic sensor with the diameter of the fiber core.

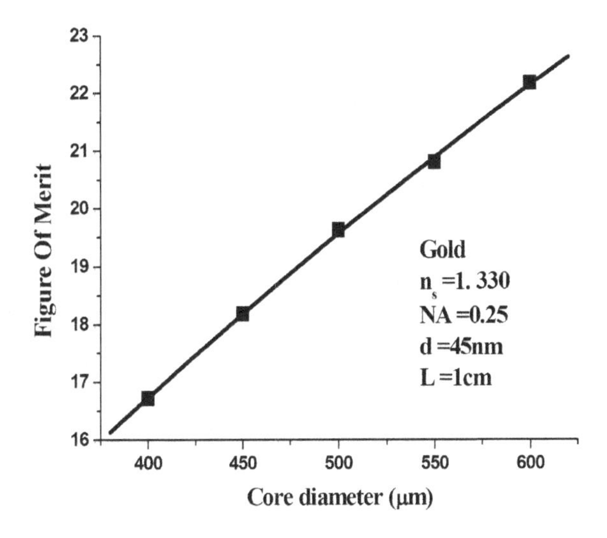

Fig. 7.3. Variation of figure of merit of the SPR based fiber optic sensor with the diameter of the fiber core.

7.2.1.2 *Sensing length*

The effect of sensing length is opposite to that of core diameter. Figures 7.4 to 7.6 show the variation of sensitivity, width of SPR

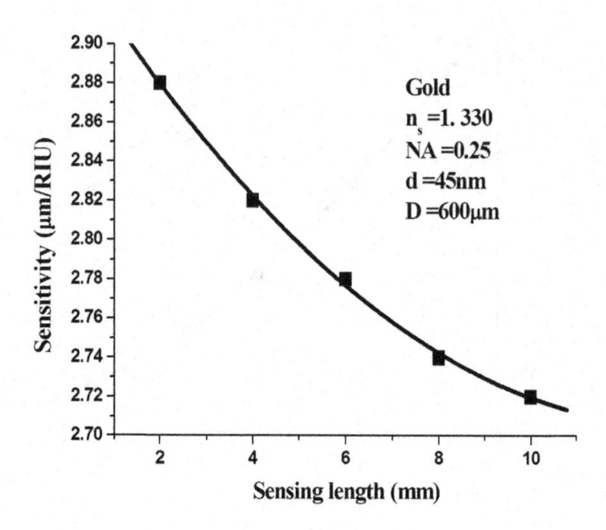

Fig. 7.4. Variation of sensitivity of the SPR based fiber optic sensor with the length of the sensing region.

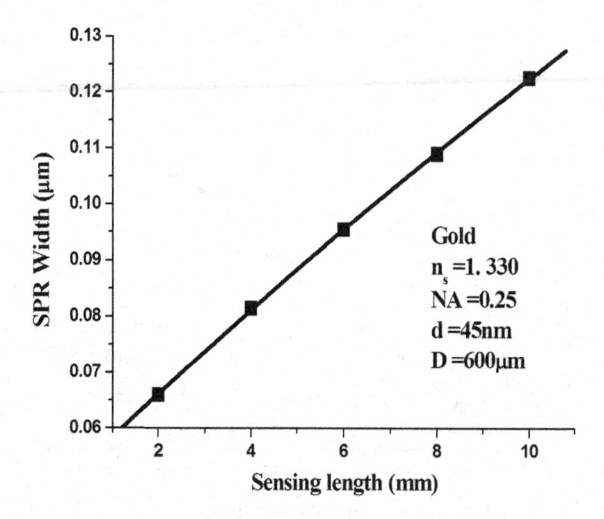

Fig. 7.5. Variation of width of the SPR spectrum of the SPR based fiber optic sensor with the length of the sensing region.

spectrum, and the figure of merit as a function of length of the sensing region, respectively. The results are plotted for 1.330 refractive index of the sensing region and 600 μm core diameter. The values of other parameters used for the simulation are the same as used for the effect

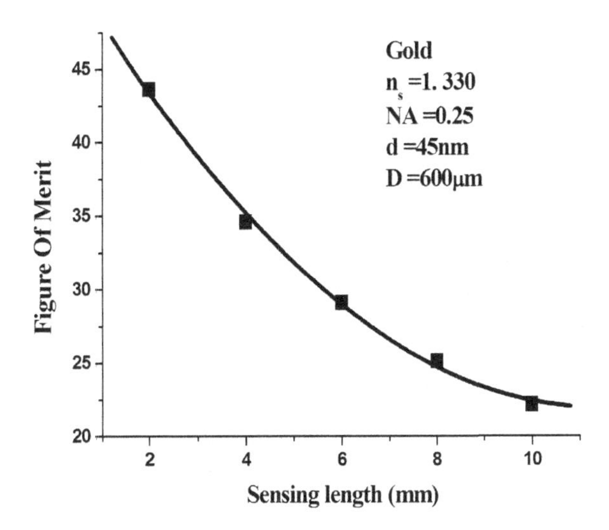

Fig. 7.6. Variation of figure of merit of the SPR based fiber optic sensor with the length of the sensing region.

of core diameter. It may be seen that as the length of the sensing region increases the sensitivity and figure of merit of the sensor decreases while the width of the SPR spectrum increases. Decrease in sensitivity and figure of merit and increase in width of SPR spectrum occur because the number of reflections of a ray in the sensing region increases as the length of the sensing region increases. The effect of increasing sensing length of the SPR probe is the same as that of decreasing the diameter of the fiber core.

7.2.1.3 *Numerical aperture*

Figures 7.7 to 7.9 show the variation of sensitivity, width of the SPR spectrum, and the figure of merit as a function of the numerical aperture of the fiber, respectively. The numerical aperture is varied from 0.17 to 0.25. The results are plotted for 1.330 refractive index of the sensing region and 600 μm core diameter. The values of other parameters used for simulation are the same as used for the effect of core diameter. The figures show that as the numerical aperture of the fiber increases the sensitivity and the spectral width of the SPR spectrum of the sensor increases, while the figure of merit

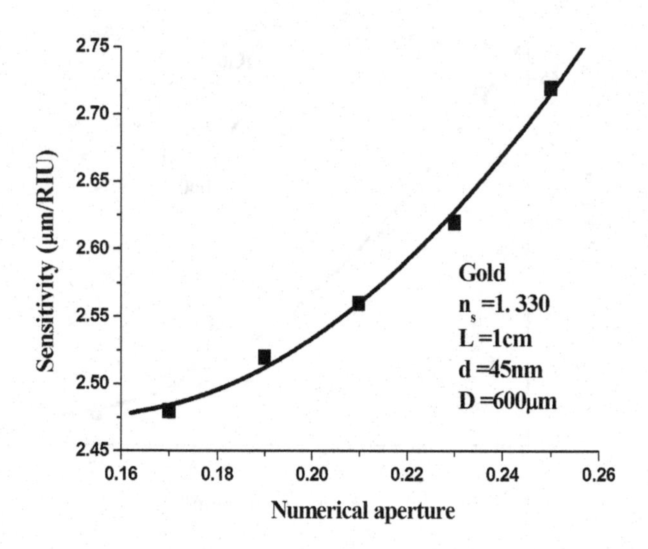

Fig. 7.7. Variation of sensitivity of the SPR based fiber optic sensor with the numerical aperture of the fiber.

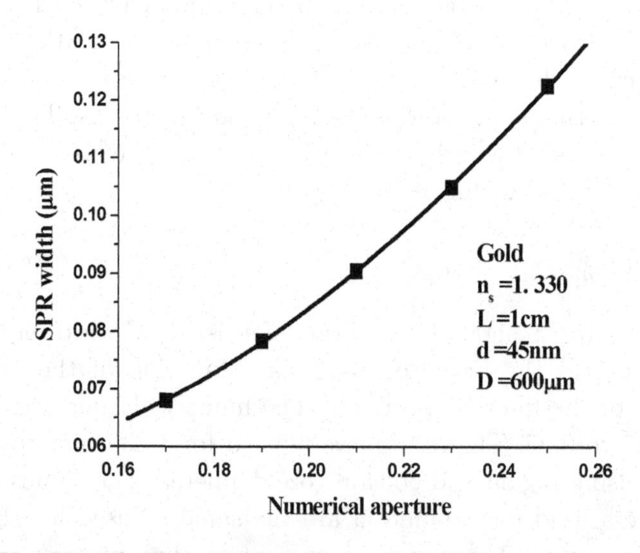

Fig. 7.8. Variation of width of SPR spectrum of the SPR based fiber optic sensor with the numerical aperture of the fiber.

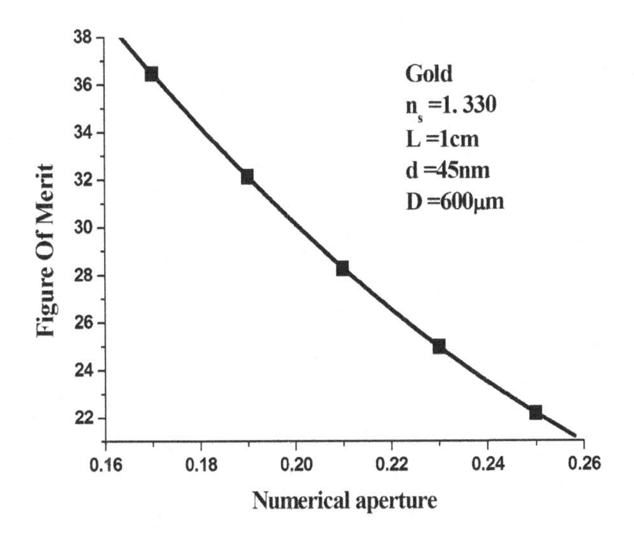

Fig. 7.9. Variation of figure of merit of the SPR based fiber optic sensor with the numerical aperture of the fiber.

of the sensor decreases. The sensitivity increases because as the numerical aperture increases the critical angle of the fiber decreases and comes more close to the critical angle of the sensing region. This is similar to reducing the angle of incidence of the ray in the sensing region by tapering the probe[2] which increases the sensitivity. When critical angle decreases the range of angles of guided rays in the fiber increases which results in the increase in the number of reflections of rays and hence increase in the width of the SPR spectrum. Thus, the width of the SPR spectrum increases with the increase in the numerical aperture of the fiber. The figure of merit which is the ratio of the sensitivity to the width of thc SPR spectrum decreases as the numerical aperture of the fiber increases.

7.2.2 *Change of metal*

If the plasmonic metal used for the sensing is changed then it will affect the resonance wavelength, the width of the SPR spectrum, and the sensitivity of the sensor for a given refractive index of the sensing medium. This occurs because each metal has different values

Table 7.1. Shift in resonance wavelength and the sensitivity of the sensor for silver, gold, and ITO coated films. The other parameters are: NA = 0.22 and fiber core diameter = 400 μm.

Metal	Optimum film thickness (nm)	$\lambda_1 (\mu m)$ ($n_s = 1.20$)	λ_2 (μm) ($n_s = 1.21$)	Shift in resonance wavelength (μm)	Sensitivity (μm/RIU)
Silver	40	0.03244	0.3313	0.0069	0.69
Gold	50	0.3831	0.3913	0.0082	0.82
ITO	70	1.3766	1.3897	0.0131	1.31

of its real and imaginary parts of the complex dielectric constant. The real part of dielectric constant is responsible for the shift in resonance wavelength while the imaginary part changes the depth of the SPR curve. Table 7.1 shows the resonance wavelength and the sensitivity of the sensor for 1.20 value of the refractive index of the sensing medium for three metals/conducting metal oxide and collimated–focusing lens combination for light launching in the fiber.[3] The values of other parameters that have been used for the simulation are given in the Table. It can be seen from the Table that the change in metal changes all these performance parameters of the sensor. The sensitivity is high in the case of indium tin oxide (ITO) while it is low for the silver coated probe. In the case of ITO based sensor the spectral width was found to be large as compared to silver and gold. The large width of the SPR spectrum is a disadvantage because it decreases the accuracy of the measurement of resonance wavelength from the SPR spectrum. The other change that can be observed is that the change in metal can be used to tune the range of resonance wavelength for the operation of the sensor.

The effect of change of metal on the sensitivity, width of the SPR spectrum, and the figure of merit of the sensor are plotted in Figs. 7.10 to 7.12 as a function of refractive index of the sensing medium, respectively. It can be seen that for a given refractive index of the sensing medium, the sensitivity of the gold coated probe is

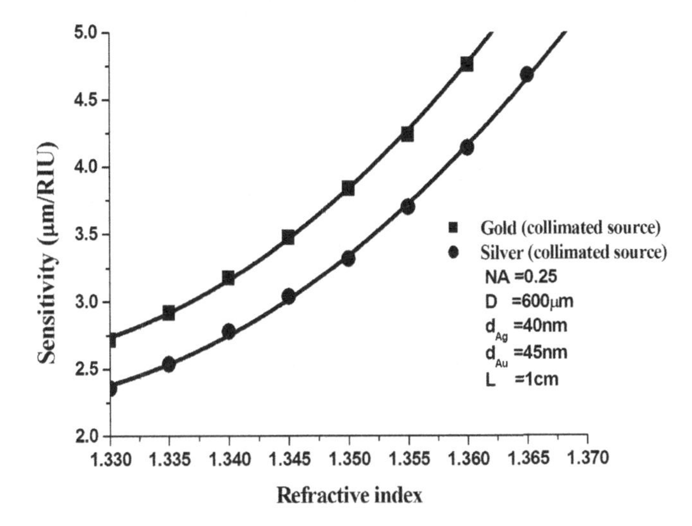

Fig. 7.10. Variation of sensitivity of the SPR based fiber optic sensor with the refractive index of the sensing medium for gold and silver coated probes.

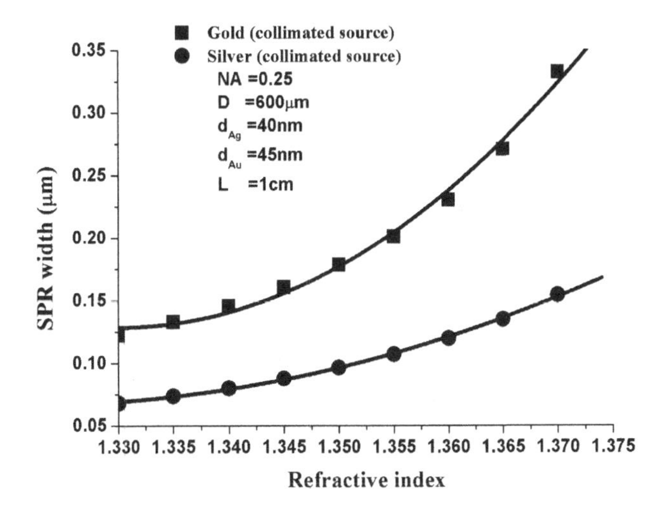

Fig. 7.11. Variation of width of the SPR spectrum of a SPR based fiber optic sensor with the refractive index of the sensing medium for gold and silver coated probes.

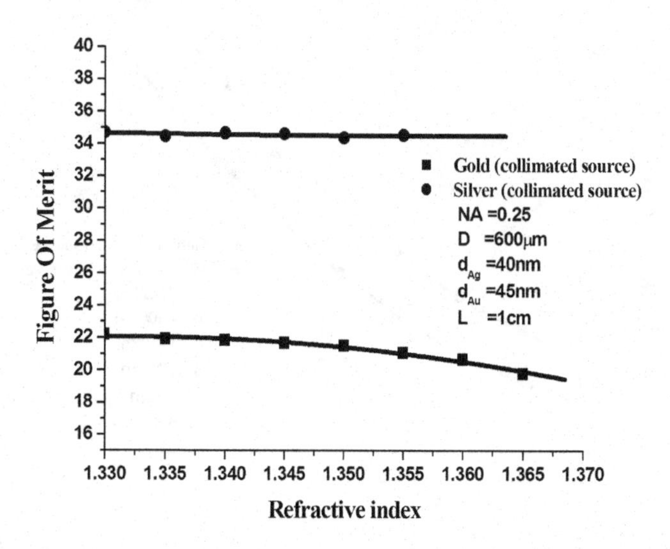

Fig. 7.12. Variation of figure of merit of the SPR based fiber optic sensor with the refractive index of the sensing medium for gold and silver coated probes.

greater than that of silver coated probe. This is because of large value of the real part of the dielectric constant of gold in comparison to silver. Thus the coating of gold is better than that of silver if high sensitivity is required.

The effect of metal on width of the SPR spectrum is opposite to that we have seen in the case of sensitivity if better performance of the sensor is required. From Fig. 7.11 it can be seen that for a given refractive index the width of the SPR spectrum is high for gold coated probe than the silver coated probe. In other words, the detection accuracy, defined in Chapter 4, is poor in the case of gold coated probe in comparison to silver coated probe. This is because the imaginary part of the dielectric constant is large in the case of silver than in the case of gold resulting in the sharper and deep SPR spectrum. Thus, silver coated SPR probe is better if high detection accuracy is required.

The change of metal gives better figure of merit for silver coated SPR probe than gold coated SPR probe as shown in Fig. 7.12. However, the effect of refractive index on figure of merit in the case of both the metals is marginal.

7.2.3 *Influence of dopants in fiber core*

Since the SPR condition, given by Eq. (2.36), depends on the refractive index of the material of the fiber core a change in the refractive index of the fiber core can affect the performance of the fiber optic SPR sensor. The refractive index of the core can be changed by using dopants in the fiber core. Generally, a pure silica fiber core is used for the SPR based fiber optic sensors. To see the effect of change in refractive index of the fiber core the performance of the SPR based fiber optic sensor was evaluated using dopants in the fiber core.[4] The dopants considered in pure silica fiber core were germanium oxide (GeO_2), boron oxide (B_2O_3), and phosphorus pent-oxide (P_2O_5). The effects of these dopants and their doping concentrations on the sensitivity of the sensor were studied theoretically.[4] The variation of sensitivity with the refractive index of the sensing medium for the five cases is shown in Fig. 7.13. It is observed that the sensitivity is highest in the case of dopant B_2O_3 (5.2) and is lowest if GeO_2 (19.3) is used. Further, as the doping concentration of GeO_2 is increased from 6.3 mole% to 19.3 mole%, a noticeable decrease in the sensitivity occurs. In other words, for a given refractive index of the sensing medium, the sensitivity decreases with the increase in the GeO_2 doping concentration.

7.2.4 *Role of high index dielectric layer*

In Sec. 6.2.1 we have discussed the SPR based fiber optic refractive index sensor using over layers of silicon and oxides over metal coated unclad fiber core. These high index dielectric layers increase the evanescent field at the interface of dielectric and sensing medium which increases the sensitivity of the sensor.[5] Here we discuss the effect of the thickness of the dielectric layer on the sensitivity of the sensor. Figure 7.14 shows the variation of sensitivity of the sensor with the thickness of the silicon layer determined from the theoretical and experimental SPR spectra obtained for different refractive indices of the sensing medium.[6] The sensitivity increases as the thickness of the silicon layer increases. The sensitivity obtained experimentally in the case of 10 nm thick silicon film is about

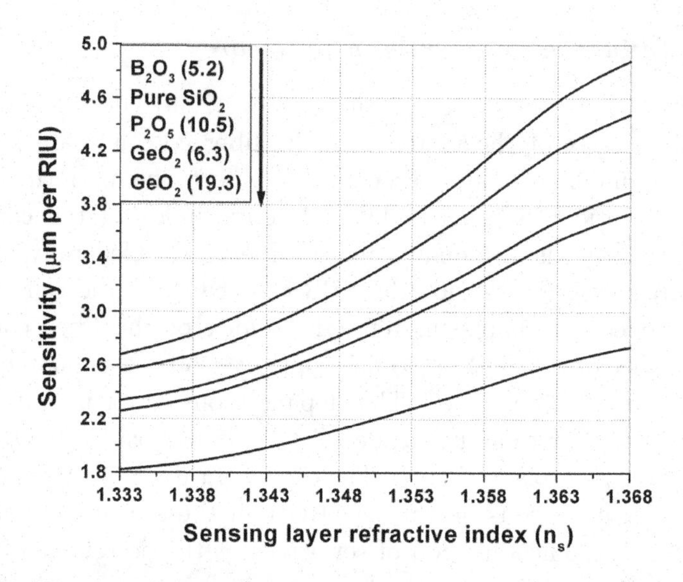

Fig. 7.13. Variation of sensitivity with sensing layer refractive index for the silica fiber core (i) doped with GeO_2 (19.3), (ii) doped with GeO_2 (6.3), (iii) doped with P_2O_5 (10.5), (iv) with no doping, and (v) doped with B_2O_3 (5.2). Numbers in brackets are the molar concentrations of dopant in mole %. Reprinted from Ref. [4] with permission from Elsevier.

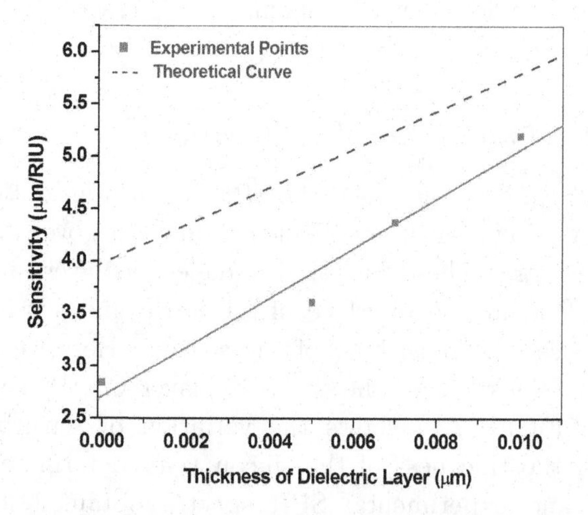

Fig. 7.14. Variation of sensitivity with the thickness of the silicon layer. Reprinted from Ref. [6] with permission from Optical Society of America.

80% greater than the probe not having silicon film. For silicon thickness greater than 10 nm the SPR spectra are very broad and their resonance wavelengths cannot be determined. Thus, the 10 nm thick film of the silicon layer can be taken as the upper limit for the sensitivity enhancement. This study suggests that the addition of a nano-meter thick film of high index dielectric over metal layer increases the sensitivity of the SPR based fiber optic sensor. There are many other high refractive index dielectric materials which can further increase the sensitivity of the sensor but their resonance wavelengths fall in the infrared region where the material of the fiber core is not very transparent.

In the study on sensitivity dependence on dielectric film thickness and the plot shown in Fig. 7.14, the n-type silicon layer was used. Later, the experiments were also performed on the probe coated with p-type silicon layer over silver coated fiber core. The variation of resonance wavelength with the refractive index of the sensing medium for the probes having over-layers of n-type and p-type silicon is shown in Fig. 7.15.[7] According to the plot, the resonance wavelength, for a given refractive index of the sensing medium, is higher in the case of probe coated with an over-layer of n-type silicon than the probe coated with p-type silicon over-layer. Further, the sensitivity of the SPR probe utilizing n-type silicon over-layer has greater sensitivity than that utilizing p-type silicon over-layer. The sensitivity values determined from the slope of the curve plotted between resonance wavelength and the refractive index of the sensing medium for silver and gold coated unclad core with silicon over-layer are tabulated in Table 7.2.

As mentioned earlier the sensitivity increases if the refractive index of the over-layer increases. The change in refractive index due to dopants (or the charge carriers) in intrinsic silicon occurs at the fourth place of decimal which is negligible if one compares the difference in the sensitivity of the sensor obtained experimentally. Thus, this does not appear to be the reason behind the difference in the sensitivity obtained between p-type and n-type silicon over-layers based fiber optic SPR sensor.

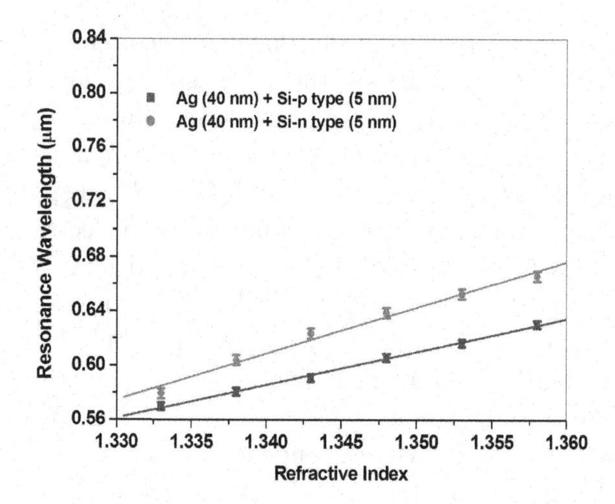

Fig. 7.15. Variation of resonance wavelength with the refractive index of the sensing medium for p- and n-type silicon over-layer coated on silver coated unclad fiber core. Reprinted from Ref. [7] with permission from Elsevier.

Table 7.2. Sensitivities of n- and p-type silicon based fiber optic SPR sensors in the case of silver and gold metal layers.[7]

Metal	Silicon (5 nm)	Sensitivity (μm/RIU)
Silver	n-type	3.3852
	p-type	2.4282
Gold	n-type	4.1102
	p-type	2.7314

The difference in sensitivity between n- and p-type silicon over-layers coated probes may be due to the difference in the charge carriers in silicon. Due to vacancies in p-type silicon, the charge density at metal–silicon interface will be higher and therefore its field probing the analyte will be sharp and hence will not be probing the analyte completely, resulting in lower sensitivity. While in case of n-type silicon, due to same type of charge carriers, there will be low charge gathering at the metal–silicon boundary so the field will be shallower in the analyte and will be probing the analyte completely and hence the sensitivity will be higher in case of n-type silicon as compared to p-type case. The other reason can be that the charge carriers in silicon may be affecting the propagation constant of the surface plasmon wave which is solely due to the collective oscillations of free electrons in metal. The holes in silicon are perturbing the propagation constant more than the free electrons in the silicon. Here we would like to mention that the sensitivity of the silver coated probe (without silicon) lies between the sensitivities of silver + p-type and silver + n-type silicon coated probes. Same is the case for the probe coated with gold only. This has suggested that both the refractive index of silicon and the sign of the charge carriers are contributing to the difference in the sensitivity of the sensor. The high refractive index of silicon increases the sensitivity while the positive charge carriers, holes, decrease the sensitivity. The effect of holes is more than the effect of silicon refractive index. In the case of n-type silicon coated probe, the sensitivity increases due to high refractive index of silicon and the negative charge carriers, electrons, resulting in sensitivity greater than that of the probe coated with silver only. In other words, negative charge carriers are increasing the sensitivity while the positive charge carriers in silicon layer are decreasing it.

In SPR sensing structure, gold or silver is generally used as a plasmonic material. However, in some studies copper and aluminium have been used as plasmonic materials. The disadvantage of most of the metals other than gold is that their coatings get easily oxidized in open atmosphere which affects the sensitivity, operational wavelength, and the stability of the sensor. Therefore, some over-layer is required which can protect the metal layer from oxidation. But it should be noted that the addition of over-layer can influence the sensitivity and the resonance wavelength of the sensor. In the aforementioned case, an additional layer of silicon over gold or silver film has been used. The addition of silicon layer introduces considerable red shift in the resonance wavelength and increases the sensitivity of the sensor. In contrast, silicon is not very chemically inert. Due to this reason, the over-layers of oxides are introduced which possess better chemical stability in open environments.

The optical fiber based SPR sensors utilizing copper and the over-layer of silicon dioxide (SiO_2), tin oxide (SnO_2), and titanium dioxide (TiO_2) have been reported.[8] The variation of resonance wavelength with the refractive index of the sensing medium for copper and TiO_2 layer of different thicknesses is shown in Fig. 7.16. In the same figure, the results for the probe coated with copper layer only are also shown. According to these experimental results, for a given refractive index of the sensing medium, the increase in the thickness of the TiO_2 layer increases the resonance wavelength implying that the resonance wavelength can be tuned by changing the thickness of the TiO_2 layer over copper coated fiber optic SPR probe. However, there is a limit to the increase in the thickness of the TiO_2 layer, because at higher thicknesses, the SPR curve gets broaden which makes the determination of resonance wavelength difficult.

The sensitivity of the sensor calculated from the slopes of the curves plotted in Fig. 7.16 for 1.34 refractive index of the sensing medium is shown in Fig. 7.17 for different thicknesses of the TiO_2 film. The sensitivity of the sensor increases with the increase in the thickness of the TiO_2 layer. The variation of sensitivity with the thickness of the TiO_2 layer determined theoretically using N-layer model matches qualitatively with the experimental curve in the sense that sensitivity increases with the increase in the thickness of TiO_2 for a given refractive index of the sensing medium. The addition of 10 nm thick TiO_2 layer over copper layer increases the sensitivity by a factor of about 2.5. The difference in sensitivity between the theoretical and the experimental results may be due to the following reasons: (i) non-inclusion of skew rays in simulation, and (ii) the theoretical model used is based on geometrical optics. The sensitivity enhancement is solely due to the addition of oxide layer of high dielectric constant over the metal layer. It also protects metal layer from oxidation, helps in tunability of operational wavelength region, and can sense gases because many oxides change their optical properties on exposure to gases.

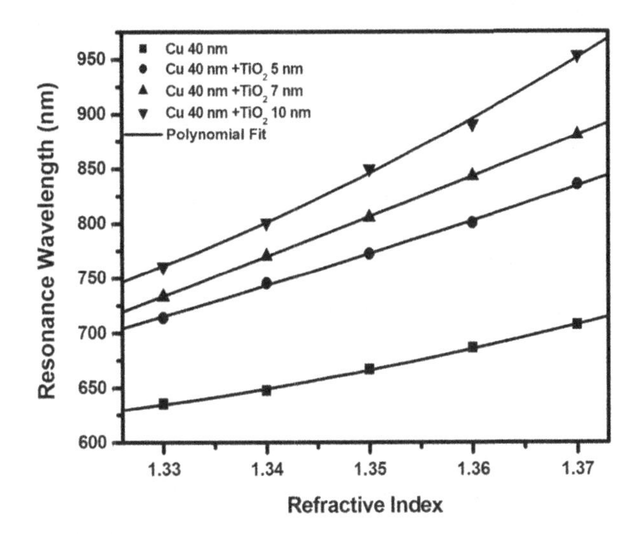

Fig. 7.16. Experimental variation of resonance wavelength with the refractive index of the sensing medium for different thicknesses of TiO_2 layer. Reprinted from Ref. [8] with permission from Elsevier.

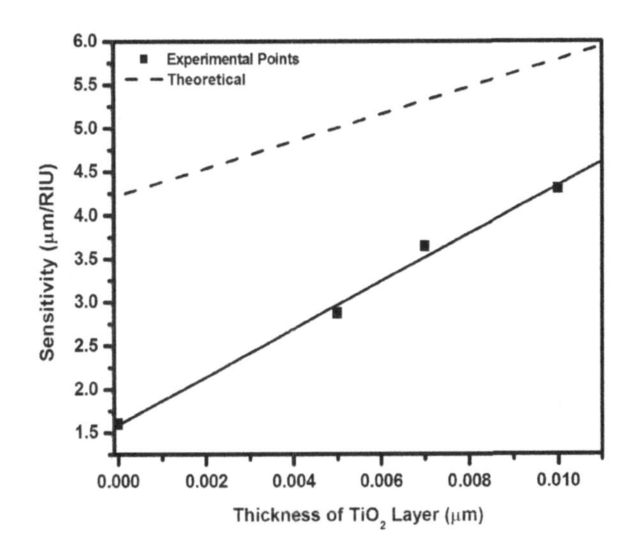

Fig. 7.17. Variation of sensitivity with the thickness of TiO_2 film for refractive index 1.34. Reprinted from Ref. [8] with permission from Elsevier.

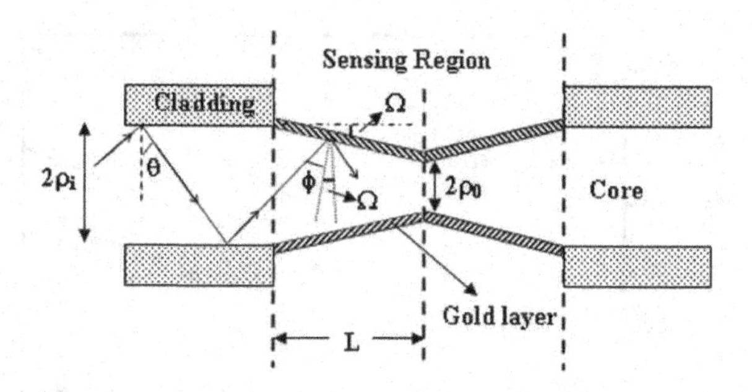

Fig. 7.18. A typical SPR based fiber optic sensor with tapered probe. Reprinted from Ref. [9] with permission from Elsevier.

7.2.5 *Probe design*

The change in probe geometry can also influence the performance of the sensor. Here we consider tapered and U-shaped probes for SPR based fiber optic sensor.

7.2.5.1 *Tapered probe*

Tapering of fiber optic probe is one of the modifications that have been carried out in the geometry of the fiber optic SPR probe. A typical SPR based tapered fiber optic probe is shown in Fig. 7.18.[9] The unclad core of the fiber is tapered and is coated with a metal layer. The guided rays in the uniform core fiber enter the sensing tapered region and their angles of incidence with the normal to the core–metal interface decrease as these rays propagate. If we assume adiabatic (slow) tapering, then the angle θ of a ray in the uniform core region transforms into angle $\phi(z)$ in the tapered region and is given by[10]

$$\phi(z) = \cos^{-1}\left[\frac{\rho_i \cos\theta}{\rho(z)}\right] - \tan^{-1}\left(\frac{\rho_i - \rho_0}{L}\right), \tag{7.1}$$

where ρ_i and ρ_0 are the radii of the taper at the input ($z = 0$) and output end ($z = L$) of the taper, respectively, $\rho(z)$ is the taper radius at a distance z from the input end of the taper and L is the length of the taper; 2 L is the length of the sensing region. The second term on right hand side of Eq. (7.1) is the taper angle (Ω). The variation of taper radius with z depends on the taper profile.

Following three taper profiles have been considered for the studies on fiber optic tapered sensor whether it is based on evanescent field absorption or SPR technique.[9,11]

Linear:

$$\rho(z) = \rho_i - \frac{z}{L}(\rho_i - \rho_0). \tag{7.2}$$

Exponential linear:

$$\rho_e(z) = (\rho_i - \rho_0)\left[e^{\left(-\frac{z}{L}\right)} - \frac{z}{L}\left(e^{-1}\right)\right] + \rho_0. \tag{7.3}$$

Parabolic:

$$\rho_P(z) = \left[\rho_i^2 - \frac{z}{L}(\rho_i^2 - \rho_0^2)\right]^{1/2}. \tag{7.4}$$

The θ_1 to θ_2 range of incident angles of the rays guided in the uniform fiber core changes to the range $\phi_1(z)$ to $\phi_2(z)$ in the taper region at a distance z due to the variation of core diameter in the tapered region. Due to this the expression of the transmitted power at the output end of the fiber given by Eq. (4.32) changes to the following expression

$$P_{trans} = \frac{\int_0^L dz \int_{\phi_1(z)}^{\phi_2(z)} R_P^{N_{ref}(\theta,z)} \frac{n_1^2 \sin\theta\cos\theta}{(1-n_1^2\cos^2\theta)^2} d\theta}{\int_0^L dz \int_{\phi_1(z)}^{\phi_2(z)} \frac{n_1^2 \sin\theta\cos\theta}{(1-n_1^2\cos^2\theta)^2} d\theta}, \tag{7.5}$$

where the values of $\phi_1(z)$ and $\phi_2(z)$ can be found by substituting the corresponding values of θ in Eq. (7.1), and N_{ref} is the total number of reflections the ray of angle θ undergoes in the tapered region of length L and radius $\rho(z)$. In the case of all guided rays in the fiber, $\phi_1(z)$ corresponds to $\theta = \sin^{-1}(n_{cl}/n_1)$ and $\phi_2(z)$ corresponds to $\theta = \pi/2$. Further, N_{ref} is given as

$$N_{ref}(\theta, z) = \frac{L}{2\rho(z)\tan(\theta + \Omega)}. \tag{7.6}$$

The calculated sensitivity of the sensor for all the guided rays for all the three taper profiles as a function of taper ratio is shown in Fig. 7.19. The sensitivity increases with the increase in the taper ratio for all the taper profiles. However, the sensitivity is maximum

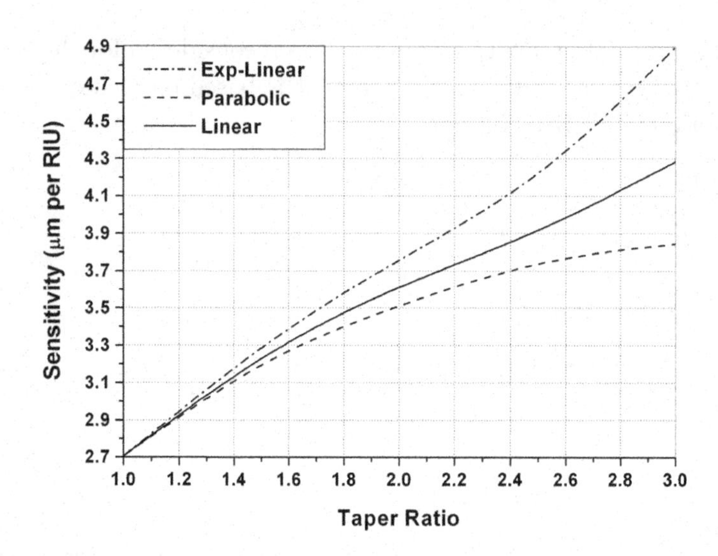

Fig. 7.19. Variation of sensitivity with taper ratio for three different taper profiles Following values of the parameters are used: NA of the fiber = 0.22, fiber core diameter = 600 μm, L = 0.5 cm, gold layer thickness = 50 nm. Reprinted from Ref. [9] with permission from Elsevier.

for exponential–linear profile (EP) while the sensitivity is minimum for parabolic profile (PP). For linear profile (LP) the sensitivity is between the two. For taper ratio 3.0, the sensitivity enhancement in the case of exponential–linear taper profile (EP) SPR sensor is more than 80% in comparison to un-tapered fiber optic SPR sensor.

To understand the effect of taper ratio and taper profiles on the sensitivity of the sensor, the analysis of the resonance condition given by Eq. (2.36) in wavelength interrogation mode is required.[9] In this equation, the propagation constant (K_{SP}) of surface plasmon wave depends on the wavelength of the light, the dielectric constant of metal layer (ε_m) and the refractive index (n_s) of the sensing medium whereas the propagation constant (K_{inc}) of the excitation light wave depends on its angle of incidence and wavelength of light in addition to the refractive index of the fiber core. Figure 7.20 shows the variation of K_{SP1} (for n_{s1} = 1.333), K_{SP2} (for n_{s2} = 1.343) and K_{inc} with wavelength for the un-tapered fiber optic SPR probe. For these plots the angle of incidence of the ray is taken as the middle

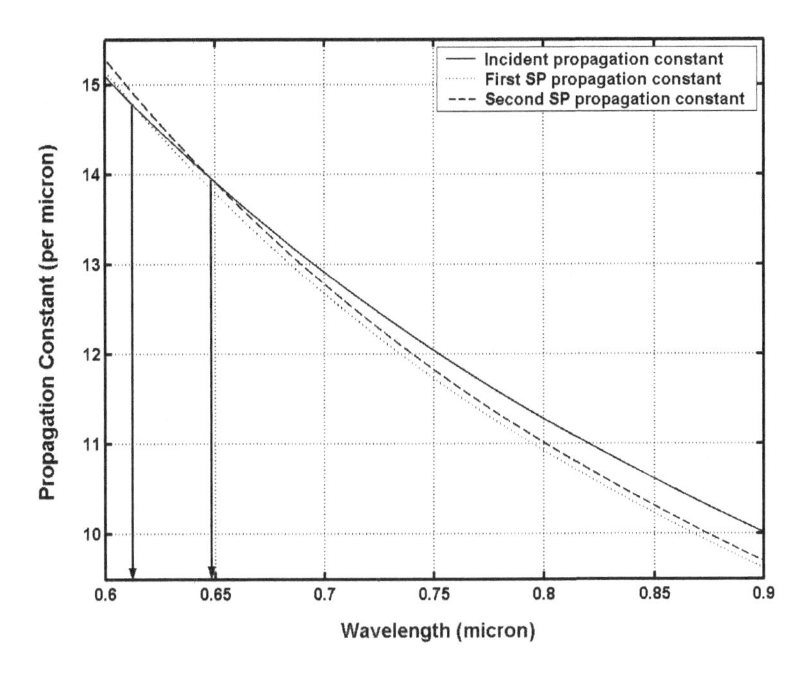

Fig. 7.20. Variation of propagation constants for SPW and incident wave with wavelength for taper ratio 1.0. Reprinted from Ref. [9] with permission from Elsevier.

value of the corresponding angle (ϕ_1, ϕ_2). For the resonance to occur the two propagation constants, K_{SP} and K_{inc}, should be equal. In Fig. 7.20, K_{SP} and K_{inc} are equal at 613.86 nm for $n_{s1} = 1.333$ and at 648.46 nm for $n_{s2} = 1.343$. Thus, a shift ($\delta\lambda_{res}$) of 34.60 nm in resonance wavelength occurs for the refractive index change (δn_s) of 0.010. Figure 7.21 shows the variation of K_{SP1}, K_{SP2}, K_{inc}(LP), K_{inc}(PP), and K_{inc}(EP) with the wavelength for taper ratio of 3.0. The K_{SP1} and K_{SP2} curves are the same curves that have been plotted in Fig. 7.20 because K_{SP} does not depend on the angle of incidence and hence on taper ratio. However, the variation of K_{inc} depends on the taper profile and the taper ratio. This is due to the following two reasons: (i) the angular range (ϕ_1, ϕ_2) is different for different taper profiles, and (ii) both ϕ_1 and ϕ_2 depend on the taper ratio. Due to this, the K_{SP1} curve intersects three K_{inc} curves at three different points and same is true for K_{SP2}

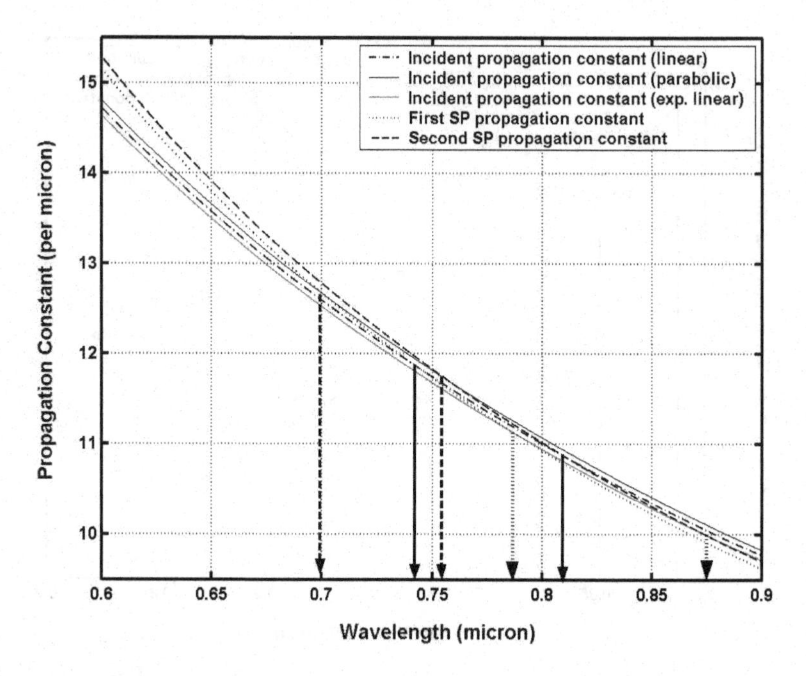

Fig. 7.21. Variation of propagation constants for SPW and incident wave with wavelength for taper ratio 3.0. Reprinted from Ref. [9] with permission from Elsevier.

curve. The intersections points corresponding to LP, PP, and EP are shown in Fig. 7.21. The wavelengths corresponding to these intersections and the corresponding shift in resonance wavelength for three taper profiles and three taper ratios are tabulated in Table 7.3. The maximum shift in resonance wavelength is in the case of EP whereas minimum shift is found for PP. For LP, the shift is between the two. This justifies the results obtained in the earlier case and shown in Fig. 7.19.

The sensitivity enhancement with taper profile and taper ratio can also be explained in terms of evanescent wave absorption. The tapering of the fiber broadens the angular range from (θ_1, θ_2) to (ϕ_1, ϕ_2). The decrease in the angle of incidence means more spreading of the evanescent field into the region adjacent to fiber core. Thus, the coupling of the evanescent wave with surface plasmon wave becomes stronger and enhances the sensitivity. For different taper profiles,

Table 7.3. Resonance wavelengths and their shift corresponding to 0.01 change in refractive index for different taper profiles and taper ratios.[9]

Taper profiles	Wavelength	Taper ratio		
		1.0	2.0	3.0
Linear	λ_1 (nm)	613.86	684.09	741.94
	λ_2 (nm)	648.46	735.56	811.59
	$\Delta\lambda$ (nm)	34.60	51.47	69.65
Parabolic	λ_1 (nm)	613.86	671.24	699.65
	λ_2 (nm)	648.46	719.23	755.58
	$\Delta\lambda$ (nm)	34.60	47.99	55.93
Exponential	λ_1 (nm)	613.86	699.57	787.58
	λ_2 (nm)	648.46	755.48	875.08
	$\Delta\lambda$ (nm)	34.60	55.91	87.50

the angular range (ϕ_1, ϕ_2) in tapered region is maximum for EP and minimum for PP. Due to this the sensitivity is maximum for EP. However the increase in the angular range due to tapered fiber broadens the SPR curve. The degree of SPR curve broadening is much smaller than the enhancement of the sensitivity.

7.2.5.2 *U-shaped probe*

In another modification, the SPR probe is bent to make it U-shaped.[12] The U-shaped SPR probe consists of an unclad bent region with a desired length of its bottom region coated with metallic layer as shown in Fig. 7.22. It should be remembered that only outer bent region is used for the SPR probe. The metallic layer on the fiber core is followed by the sensing medium. Since inner bent region is not coated with metal, its use is to guide light in this region by means of total internal reflection. The probe shown in Fig. 7.22 is not to the scale.

To analyze this kind of probe, 2-dimensional approach was considered. Further, all the guided rays and their electric vectors (because p-polarized light excites surface plasmons) were assumed to be confined in the plane of bending. If θ is the angle of incidence of a ray in the straight region of the U-shaped probe then it transforms

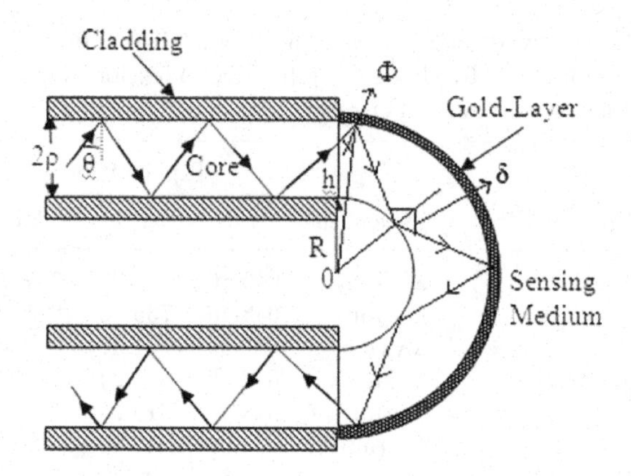

Fig. 7.22. A typical U-shaped fiber optic SPR probe (not to scale). Reprinted from Ref. [12] with permission from Institute of Physics (IOP).

to the angles

$$\phi(h) = \sin^{-1}\left[\frac{(R+h)}{(R+2\rho)}\sin\theta\right], \tag{7.7}$$

and

$$\delta(h) = \sin^{-1}\left[\frac{(R+h)}{R}\sin\theta\right], \tag{7.8}$$

at the outer and the inner surfaces of the U-region respectively. In Eqs. (7.7) and (7.8), h is the distance of the point of intersection of the ray at the entrance face of the bending region from the core–cladding interface of the inner surface of the U-region; R is the bending radius of the inner circular region and ρ is the core radius of the fiber.

For θ_1 to θ_2 range of incident angles of the guided rays in the straight region of the U-shaped probe, the transmitted power received at the other end of the U-shaped probe is given by the following expression similar to that written for tapered probe (Eq. (7.5))

$$P_{trans} = \frac{\int_0^{2\rho} dh \int_{\phi_1(h)}^{\phi_2(h)} R_P^{N_{ref}} \frac{n_1^2 \sin\theta\cos\theta}{(1-n_1^2\cos^2\theta)^2} d\theta}{\int_0^{2\rho} dh \int_{\phi_1(h)}^{\phi_2(h)} \frac{n_1^2 \sin\theta\cos\theta}{(1-n_1^2\cos^2\theta)^2} d\theta}, \tag{7.9}$$

where R_P is the reflection coefficient for the SPR sensor geometry, ϕ_1 and ϕ_2 are the angles of incidence of the rays with the outer surface of the U-shaped probe corresponding to angles θ_1 and θ_2 and $N_{ref}(\theta)$ is the total number of reflections made by the ray at the outer surface of the U-shaped probe for the ray making an angle θ with the normal to the core–cladding interface and is given as

$$N_{ref} = \frac{L}{8\rho}\left[\cot\theta + \cot\left(\left(\frac{R+2\rho}{R}\right)\theta\right)\right]. \qquad (7.10)$$

In Eq. (7.10), L is the length of the metal coated sensing region of the probe. For all guided rays in the straight region of the probe, ϕ_1 corresponds to $\theta = \sin^{-1}(n_{cl}/n_1)$ and ϕ_2 corresponds to $\theta = \pi/2$; n_1 and n_{cl} are the refractive indices of the fiber core and cladding respectively.

Figure 7.23 shows the variation of calculated sensitivity of the sensor with the bending radius of the U-shaped probe for three different values of sensing length.[12] The sensitivity first increases as the bending radius decreases and after attaining some maximum value it starts decreasing. The maximum sensitivity occurs at a particular bending radius. For bending radius lower than this, the sensitivity decreases due to the angle of incidence of the last ray being less than the angle of incidence required for the ray to be guided (critical angle). The variation is nearly independent of the sensing length because the sensing length does not affect the angle of incidence. The maximum sensitivity of the sensor is obtained for 1 cm bending radius and 3 cm length of the sensing region of the probe and its values is about 25 times greater than that of the conventional straight fiber SPR probe. The sensitivity of the U-shaped probe increases with an increase in the sensing length because increase in sensing length increases the number of reflections in the sensing region and hence more evanescent field spreads into the sensing region. The sensitivity enhancement mainly occurs due to the increase in the penetration depth of the evanescent field. As the bending increases, the angle of incidence decreases and comes close to the critical angle which increases the penetration depth of the evanescent field in the sensing region. This increases the coupling of the evanescent wave and the surface plasmon wave resulting in the increase in the sensitivity of the sensor.

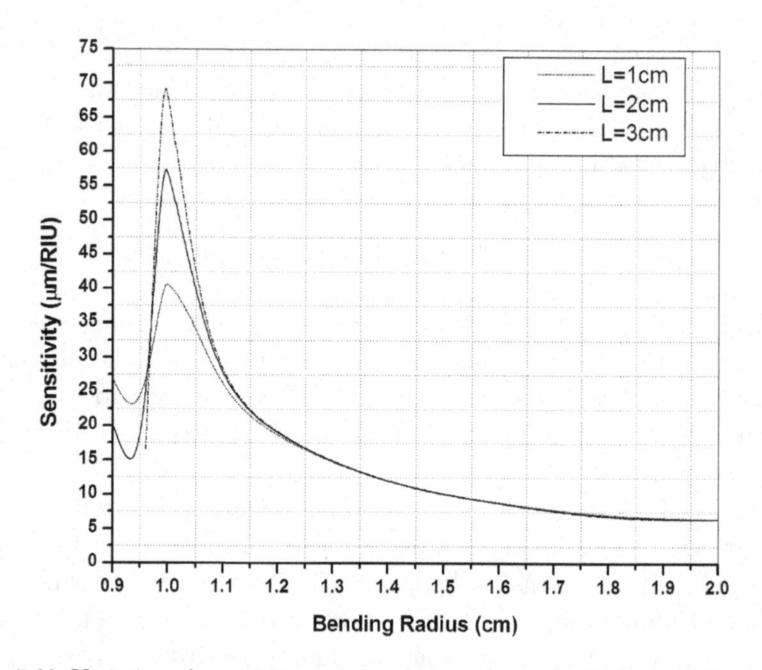

Fig. 7.23. Variation of sensitivity with bending radius of the U-shaped SPR probe for three different values of the sensing length. Reprinted from Ref. [12] with permission from Institute of Physics (IOP).

Similar to the tapered probe the enhancement of the sensitivity can also be explained in the case of U-shaped probe using propagation constants of incident and surface plasmon waves. Figure 7.24 shows the variation of propagation constant of the incident wave with the wavelength for three different values of the bending radius of the probe. In the same figure, the variation of propagation constant of the surface plasmon wave for two different values of the refractive index of the sensing region is also plotted. For the calculation of K_{inc}, the angle of incidence considered was the average value of the angles of incidence in the sensing region (ϕ_1, ϕ_2). The three propagation constants of incident wave K_{inc1} (R $=$ 0.95 cm), K_{inc2}(R $=$ 2.0 cm) and K_{inc3} (R $=$ 5.0 cm) intersect the two surface plasmon wave propagation constants K_{SP1} (for $n_s = 1.333$) and K_{SP2}($n_s = 1.335$) at six different points. According to these points of intersection, the shifts in resonance wavelength for three different bending radii of the probe are mentioned in Fig. 7.24. The shift in the resonance

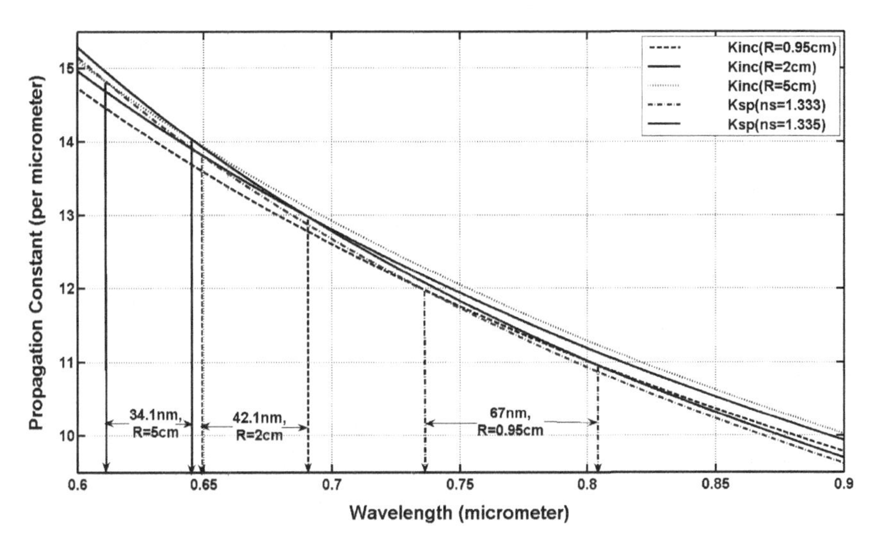

Fig. 7.24. Variation of propagation constants of surface plasmon wave and incident wave with wavelength. Reprinted from Ref. [12] with permission from Institute of Physics (IOP).

wavelength is minimum for the probe with bending radius 5 cm and it is maximum for the probe with bending radius 0.95 cm. The shift in resonance wavelength for the probe with bending radius 2 cm is between these two probes. These findings imply that the sensitivity of the U-shaped probe increases with the decrease in the bending radius of the probe.

7.3 Influence of Extrinsic Stimuli

The performance of the SPR based fiber optic sensors can be affected by external stimuli such as temperature of the environment, presence of ions in the fluid sample, etc. Here we shall learn the influence of temperature and presence of ions in the fluid on the calibration curve of the SPR based fiber optic sensor.

7.3.1 *Influence of temperature*

The change in temperature of environment can affect the dielectric properties of the fiber core and the metal used for the coating.

A study on the effect of temperature on the performance of the SPR based fiber optic sensor has been reported.[13] For the temperature dependence of refractive index of the fiber core the thermo optic coefficient (dn/dT) of silica was used along with Sellmeier relation given by Eq. (4.1). The value of thermo optic coefficient used was 1.28×10^{-5} per K. The dielectric constant of the metal depends on the plasma and collision frequencies according to Drude model and is given by Eq. (4.2). Both these frequencies depend on temperature. The plasma frequency varies due to volumetric effect and its temperature dependence is expressed as

$$\omega_p = \omega_{p0}[1 + \gamma_e(T - T_0)]^{-1/2}, \tag{7.11}$$

where γ_e is the thermal volume expansion coefficient of the metal and T_0 is the room temperature which acts as the reference temperature. The temperature dependence of collision frequency is the sum of the collision frequencies due to phonon–electron scattering (ω_{cp}) and due to electron–electron scattering (ω_{ce}). Thus, the collision frequency is written as

$$\omega_c = \omega_{cp} + \omega_{ce}. \tag{7.12}$$

The temperature dependence of ω_{cp} is written as

$$\omega_{cp}(T) = \omega_0 \left[\frac{2}{5} + 4\left(\frac{T}{T_D}\right)^5 \int_0^{T_D/T} \frac{z^4 dz}{e^z - 1}\right]. \tag{7.13}$$

where T_D is the Debye temperature and ω_0 is a constant. The derivation of ω_{cp} is based on the simple Debye model for the phonon spectrum and is valid for $E_F >> K_B T, K_B T_D$, where E_F is the Fermi energy. For $E_F << K_B T$, $K_B T_D$, the aforementioned expression becomes

$$\begin{aligned}\omega_{cp}(T, \omega \to 0) &= \frac{\omega_p^2}{4\pi\sigma(0)} \\ &= \omega_0 \left[4\left(\frac{T}{T_D}\right)^5 \int_0^{T_D/T} \frac{z^5 dz}{(e^z - 1)(1 - e^{-z})}\right]. \end{aligned} \tag{7.14}$$

The temperature dependence of ω_{ce} can be written as

$$\omega_{ce}(T) = \frac{1}{6}\pi^4 \frac{\Gamma\Delta}{hE_F}\left[(k_B T)^2 + \left(\frac{h\omega}{4\pi^2}\right)^2\right], \qquad (7.15)$$

where Γ is a constant and Δ is the fractional Umklapp scattering. Apart from dielectric constant, the thermal expansion of metal film is also important. Since the metal film may expand only into the normal direction, the corrected thermal expansion coefficient (α') for the expansion of the metal film thickness is used which is expressed as

$$\alpha' = \alpha\frac{(1+\mu)}{(1-\mu)}. \qquad (7.16)$$

Where α is the linear thermal expansion coefficient of the bulk material and μ is Poisson number of the metal. The value of μ is close to 0.3 and thus the value of 'α' is almost twice the usual thermal expansion coefficient (α). The values of the metal parameters used for numerical simulations are the following: fiber core diameter = 600 μm, NA of the fiber = 0.25, metal film thickness = 50 nm and refractive index of sensing medium = 1.333. Figure 7.25 shows SPR curves at different temperatures.[13] Increase in the temperature shifts the SPR curves to blue side or decreases the resonance wavelength. This effect is also used for temperature sensing.

7.3.2 *Influence of ions*

In this section we discuss the effect of ionic and non-ionic solutions on the SPR spectrum of a fiber optic SPR sensor.[14] For this study, four ionic (sodium chloride [NaCl], potassium chloride [KCl], magnesium chloride [$MgCl_2$], and sodium suplhate [Na_2SO_4]) and two non-ionic (sucrose and urea) compounds were used for the preparation of solutions in Millipore water with refractive index ranging from 1.335 to 1.370.[14] The variation of resonance wavelength with the refractive index of the solution for all the compounds is shown in Fig. 7.26. As expected the resonance wavelength increases non-linearly with the increase in the refractive index of the solution.

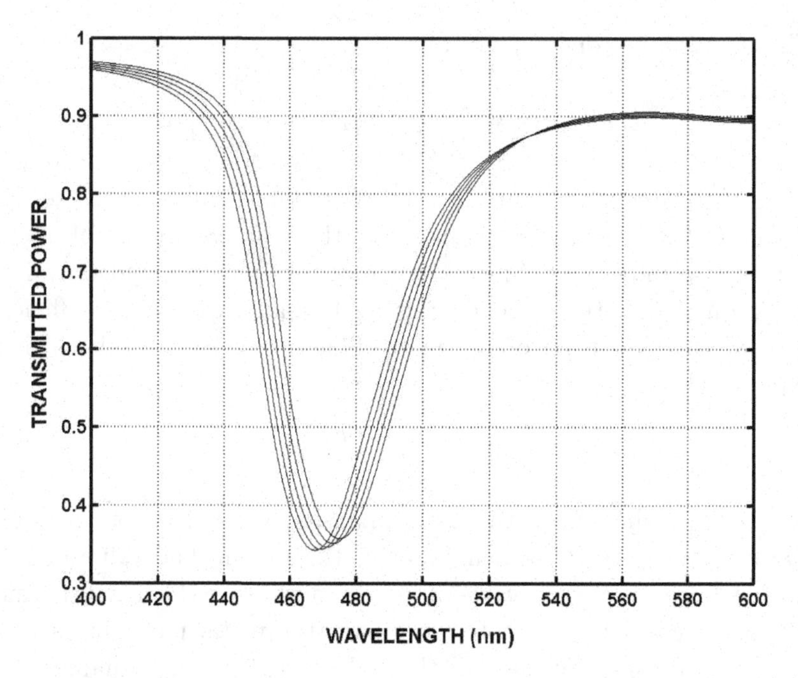

Fig. 7.25. SPR curves for 1.333 refractive index and different temperatures of the sensing medium. The curves are for 300, 375, 450, 525, and 600 K from right to left. Reprinted from Ref. [13] with permission from Optical Society of America.

However, for a given refractive index, the resonance wavelength is not same for all compounds. It is greater for ionic solutions than for non-ionic solutions. Further, the resonance wavelength is nearly the same in the case of sucrose and urea for the same refractive index, but it is different in the case of ionic solutions and the difference increases as the refractive index of the ionic solution increases. This implies that the presence of ions affects the resonance wavelength or the resonance condition. Increase in the refractive index of the ionic solution increases the ionic content in the solution which increases the resonance wavelength. The differential increase in resonance wavelength in the case of ionic fluids has been reported to be due to the electrostatic interactions of the free ions with the surface electrons of the metal film resulting in the decrease in the wave vector of surface plasmons and an increase in the

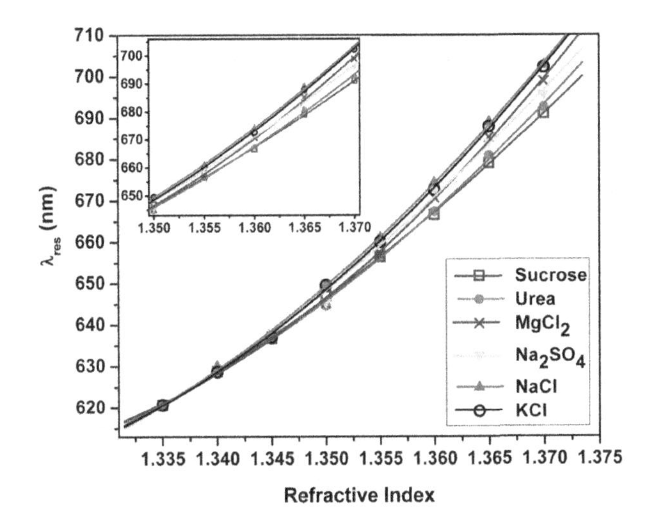

Fig. 7.26. Variation of resonance wavelength (λ_{res}) with the refractive index of sucrose, urea, Na_2SO_4, $MgCl_2$, NaCl, and KCl solutions. Reprinted from Ref. [14] with permission from Elsevier.

resonance wavelength. Further, the resonance wavelength plots of NaCl and KCl solutions overlap due to nearly the same ionic content in NaCl and KCl solutions. The additional increase in resonance wavelength, in the case of $MgCl_2$ and Na_2SO_4, is less than that obtained for NaCl and KCl solutions of the same refractive index which is due to more ionic character of NaCl and KCl than $MgCl_2$ and Na_2SO_4.

To support this, the variation of ion concentration with the refractive index of the solutions of ionic compounds is shown in Fig. 7.27. The ion concentration increases in a similar way as the resonance wavelength in Fig. 7.26 with the increase in the refractive index of all the ionic solutions. For a given refractive index, the ion concentrations of NaCl and KCl are nearly the same as also for Na_2SO_4 and $MgCl_2$. However, it is greater for NaCl and KCl than for Na_2SO_4 and $MgCl_2$. Thus, the resonance wavelength is greater in the case of NaCl and KCl than in the case of Na_2SO_4 and $MgCl_2$ which is in accordance with Fig. 7.26. It was also confirmed that the change in ionic conductivity of the samples is not responsible for additional increase in the resonance wavelength.

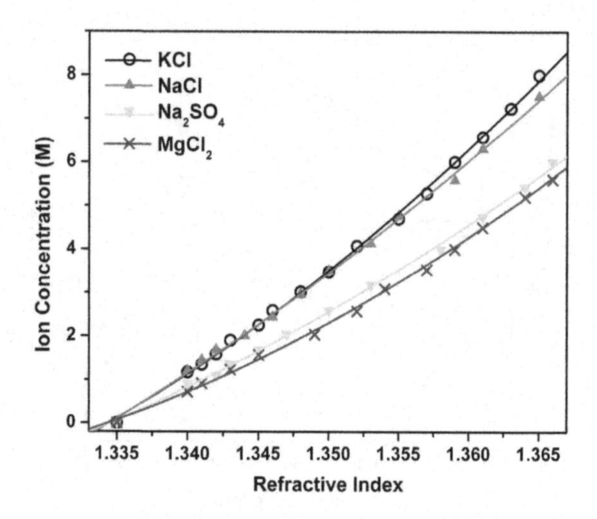

Fig. 7.27. Variation of ion concentration with the refractive index of sucrose, urea, Na_2SO_4, $MgCl_2$, NaCl, and KCl solutions. Reprinted from Ref. [14] with permission from Elsevier.

This study implies that the calibration curve of the sensor for non-ionic solutions cannot be used for ionic solutions. Further, the calibration curve differs from one ionic solution to the other ionic solution. Thus, the findings of this study are more important when the ions are produced in the solution during some chemical reaction. Therefore, the effect of ions must be incorporated as a correction term otherwise the results may get misinterpreted.

7.4 Summary

In this chapter we have presented the effects of various intrinsic and extrinsic stimuli on the performance of the SPR based fiber optic sensor. The intrinsic stimuli include fiber parameters, change of metal, dopants in fiber core, addition of high refractive index dielectric layer and the modification of SPR probe while extrinsic stimuli considered are temperature and presence of ions in the fluid sample. These studies are helpful for the designing of a SPR based fiber optic sensor.

References

1. S. Singh, R.K. Verma and B.D. Gupta, LED based fiber optic surface plasmon resonance sensor, *Optical and Quantum Electronics* **42** (2010) 15–28.
2. R.K. Verma, A.K. Sharma and B.D. Gupta, Modeling of tapered fiber-optic surface plasmon resonance sensor with enhanced sensitivity, *IEEE Photonics Technology Letters* **19** (2007) 1786–1788.
3. R.K. Verma and B.D. Gupta, Surface plasmon resonance based fibre optic sensor for the IR region using a conducting metal oxide film, *Journal of Optical Society of America A* **27** (2010) 846–851.
4. A.K. Sharma, Rajan and B.D. Gupta, Influence of different dopants on the performance of a fiber optic SPR sensor, *Optics Communications* **274** (2007) 320–326.
5. A. Lahav, M. Auslender and I. Abdulhalim, Sensitivity enhancement of guided-wave surface plasmon resonance sensors, *Optics Letters* **33** (2008) 2539–2541.
6. P. Bhatia and B.D. Gupta, Surface-plasmon-resonance-based fiber-optic refractive index sensor; sensitivity enhancement, *Applied Optics* **50** (2011) 2032–2036.
7. P. Bhatia and B.D. Gupta, Surface plasmon resonance based fiber optic refractive index sensor utilizing silicon layer: effect of doping, *Optics Communications* **286** (2013) 171–175.
8. S. Singh, S.K. Mishra and B.D. Gupta, Sensitivity enhancement of a surface plasmon resonance based fibre optic refractive index sensor utilizing an additional layer of oxides, *Sensors and Actuators A* **193** (2013) 136–140.
9. R.K. Verma, A.K. Sharma and B.D. Gupta, Surface plasmon resonance based tapered fiber optic sensor with different taper profiles, *Optics Communications* **281** (2008) 1486–1491.
10. B.D. Gupta, A. Sharma and C.D. Singh, Evanescent wave absorption sensors based on uniform and tapered fibers: a comparative study of their sensitivities, *International Journal of Optoelectronics* **8** (1993) 409–418.
11. B.D. Gupta and C.D. Singh, Fiber optic evanescent field absorption sensor: a theoretical evaluation, *Fiber and Integrated Optics* **13** (1994) 433–443.
12. R.K. Verma and B.D. Gupta, Theoretical modelling of a bi-dimensional U-shaped surface plasmon resonance based fibre optic sensor for sensitivity enhancement, *Journal of Physics D:Applied Physics* **41** (2008) 095106.
13. A.K. Sharma and B.D. Gupta, Influence of temperature on the sensitivity and signal-to-noise ratio of a fiber optic surface plasmon resonance sensor, *Applied Optics* **45** (2006) 151–161.
14. S.K. Srivastava and B.D. Gupta, Effect of ions on surface plasmon resonance based fiber optic sensor, *Sensors and Actuators B* **156** (2011) 559–562.

Chapter 8

Future Scope of Research

This book gives a rigorous discussion of the history, physics, designing, fabrication, and characterization characteristics of the sensors based on surface plasmon resonance (SPR). The chronological development of the field was given for the understanding of the development of the SPR based fiber optic sensors. Researchers have worked on various aspects of these sensors, such as enhancement of sensitivity, specificity, detection accuracy, etc. As it was observed that the performance of such sensors depends on the designing of the fiber optic probe, a number of studies were performed to check the effect of various parameters of optical fibers such as dopants/doping concentration in the core, numerical aperture, core diameter, material of the optical fiber, and light launching conditions. Many of these studies are purely theoretical and are based on certain assumptions. There is a room for plenty of work to be done in optimizing these parameters for SPR based sensing applications. Further, we have discussed various kinds of modulation techniques for SPR based sensing, such as long-range surface plasmon resonance (LRSPR), near-guided-wave SPR (NGWSPR) etc. Efforts are required for the optimization of these kinds of configurations on optical fibers. Further, sensors for the detection of hazardous chemicals can be made on optical fibers. This leaves plenty of opportunities for researchers and industrialists to work on real time, online, and remote sensing using these sensors. In most of the studies using these fibers, the sensing region was chosen to be large, which limits the opportunity for use of these sensors for small quantities of analytes in critical conditions, such as blood detection. This issue opens wide research opportunities for the enhancement of performance with respect to the analyte volume. Further, it was obtained that different metals and their combinations can be used for SPR in different spectral regions. This leads to fabrication of multi-analyte sensors by using a combination of different metals whose SPR curves are fairly separated. A biosensor using gold and silver metals was reported for simultaneous detection of glucose and urea. Optical fibers pose a wonderful opportunity for fabrication of multiple sensing regions on a single optical fiber. However, since a number of polychromatic guided rays excite multiple SPRs, the overall characteristic SPR curve is broadened. The broad nature of fiber optic SPR curves imposes a constraint over the fabrication of sensors for the analysis of four–five analytes together. However, a recent theoretical modeling study showed that if carefully fabricated, three analytes can simultaneously be detected in visible region of the electromagnetic spectrum by multi-analyte sensors fabricated on a single optical fiber probe. We hope that further ideas for improving the multi-analyte sensing using optical fibers will come forward to detect more than three analytes simultaneously on a single optical fiber. Other interesting field of SPR based fiber optic sensing may be the optimization studies of various immobilization methods and enhancing the performance. Different research groups have detected similar analytes using different immobilization techniques. A platform for the discussion of the best possible immobilization method for the detection of a particular analyte is also much required. Various immobilization strategies utilizing fiber optic SPR probes can be employed for fabrication of miniature sensors for simultaneous online monitoring of analytes of medical/biomedical importance. It may revolutionize the medical diagnostics in hospitals, where fiber optic sensors can be used for online monitoring of various blood/urine analytes from a remote central control room. Again, since the fiber optic SPR sensors are in early stage of research and they still need to be commercialized for field applications, it is also possible to integrate certain plasmonic sensor chips with single mode optical fibers, which are further miniature and require less sample volumes. Various designs such as tapers, multi-tapers, and U-shaped sensor probes were proposed for enhancement of the sensitivity of these sensors. U-shaped probes may be quite useful for point sensing, which requires very small volumes of the sample. However, no experimental study yet has been reported for such a sensor. It also opens possibilities for fiber optic U-shaped point sensors with increased sensitivity. Further, though tapered sensors have higher sensitivity, they are prone to breakage. A multi-taper design overcomes this difficulty, but still suffers with poor detection accuracies due to widening of the SPR curves with tapering. Efforts might be made in the directions of improving the detection accuracy. Most of the studies on fiber optic SPR sensors have been performed on blood analytes, where one does not need very high sensitivity, but relatively large dynamical range of the sensor to cover the whole physiological range. However, when it comes to detect pathogens or potentially harmful compounds, the fiber probes with enhanced sensitivity need to be used. These sensors might be very useful for field applications, such as detecting pollutants in drinking water such as bacteria or endocrine disruptors, or trace poisonous metals. This option also leaves a lot of opportunities to work on the enhancement of sensitivity of such sensors. The modeling and fabrication of multi-tapered probes with different taper profiles can be carried out for further enhancement of sensitivity and detection accuracy. An advantage of fiber optic sensors is that they are immune to electrical as well as electromagnetic interferences. However, it is possible that false signals may get generated due to non-specific bindings over the functionalized surfaces. Though a number of methods, in general, have been proposed to stop non-specific bindings, various methods show limitations to their areas of application, environmental conditions, and protocols. Lots of research focus is still required to work on the efficient sensing protocols which avoid the non-specific bindings, last for longer times and are robust to environmental conditions. The performance of the fiber optic SPR sensors may also get affected from ambient conditions. Simulations and experiments regarding the influence of factors such as potential, magnetic field, temperature, and pressure, etc., which may affect the SPR condition, might be an important topic of research, as its applications in various devices and ambient conditions requires a complete knowledge and control of factors affecting it.

Appendix A

Dispersion Relations of Dielectric Materials and Metals

A.1 Optical Absorption

The basis of frequency dependence of any dielectric material can be understood by analyzing the interaction of an electromagnetic wave with the array of atoms constituting that material. An atom may respond to an incoming light in two different ways, depending on the incident frequency (and, therefore, on the incoming photon energy). Generally, the atom "scatters" the light, redirecting it without altering it otherwise. On the other hand, atom absorbs the light and transits to the corresponding higher energy level in the condition of the coinciding of the photon's energy with the energy of one of the excited states. This process of lifting of the photon by the dielectric material is known as optical absorption. All material media take part in optical absorption at their corresponding different frequencies.

A.2 Dispersion Relations: Dielectrics and Metals

The dependence of the dielectric constant of a material on the frequency of the incident radiation is a known fact. Number of models have been reported in literature to describe this relationship. In this section, we shall describe models for dielectrics and metals.

A.2.1 *Dielectrics: Lorentz model of damped oscillators*

Let us consider a medium as an assembly of a large number of polarizable atoms, very small in size and closely packed to its neighbors. Upon the incidence of an electromagnetic wave on such a material, each atom acts as a classical forced oscillator which is being driven by the time-varying electric field $E(t)$ of the electromagnetic wave in any arbitrary x-direction. The incident field exerts force (F_E) on an electron and is given as:

$$F_E = eE(t) = eE_o \cos \omega t, \tag{A.1}$$

where ω and E_o, are the frequency and amplitude of the field respectively and e is the charge of an electron. Further, a restoring force exists in the direction opposite to this driving force. The Newton's equation of motion for this system may be given as:

$$eE_o \cos \omega t - m_e \omega_o^2 x = m_e \frac{d^2 x}{dt^2}, \tag{A.2}$$

The first term on the left hand side of Eq. (A.2) is the driving force, and the second one is the restoring force in the direction opposite to that of the driving force. Further, the restoring force can be supposed to have a form $F = -k_E x$, where k_E is elastic constant like a spring constant. Let us imagine a momentarily disturbed electron bound in this way oscillating about its equilibrium position with a natural or resonance frequency given by $\omega_0 = \sqrt{k_E/m_e}$, where m_e is the mass of an electron. ω_o is the oscillatory frequency of the un-driven system. Further, we have considered that the electron oscillates at the same frequency as the electric field $E(t)$. Therefore, a solution for the above equation may be considered as:

$$x(t) = x_o \cos \omega t. \tag{A.3}$$

The substitution of $x(t)$ in Eq. (A.2) gives us the expression:

$$x(t) = \frac{e}{m_e(\omega_o^2 - \omega^2)} E_o \cos \omega t, \tag{A.4}$$

or

$$x(t) = \frac{e}{m_e(\omega_o^2 - \omega^2)} E(t), \tag{A.5}$$

Equation (A.5) gives the relative displacement between the negative electron cloud and the positive nucleus. Without any incident wave, the oscillator will vibrate at its resonance frequency ω_o. In the presence of a field with a frequency less than resonance frequency ($\omega < \omega_o$), $E(t)$ and $x(t)$ will have the same sign, which implies that the oscillator can follow the applied force (in-phase oscillation). On the other hand, when $\omega > \omega_o$, the displacement will be in the direction opposite to that of the applied field (out-of-phase oscillation).

The dipole moment of the oscillating electron, a product of charge and displacement, is equal to ex. If N_e is the number of electrons per unit volume, the electric polarization is given as:

$$P = exN_e, \tag{A.6}$$

Use of Eq. (A.5) in (A.6) gives:

$$P = \frac{e^2 N_e E(t)}{m_e(\omega_o^2 - \omega^2)}, \tag{A.7}$$

For most of the materials, electric polarization P and incident field E are proportional and can be related as:

$$(K - K_o)E = P, \tag{A.8}$$

or,

$$K = K_o + \frac{P(t)}{E(t)}, \tag{A.9}$$

where K_0 is the permittivity of free space and K is called the permittivity of the material. Substitution of Eq. (A.7) in Eq. (A.9) gives the dimensionless quantity:

$$\varepsilon(\omega) = \frac{K}{K_o} = 1 + \frac{e^2 N_e}{\varepsilon_o m_e(\omega_o^2 - \omega^2)}, \tag{A.10}$$

which is called the dielectric constant of the material. Equation (A.10) is known as the dispersion relation for the frequency dependent dielectric constant [$\varepsilon(\omega)$] of any material. Further, if we include a damping force, proportional to the speed, of the form

$m_e \gamma dx/dt$ in the equation of motion (Eq. (A.2)), the dispersion equation would become:

$$\varepsilon(\omega) = 1 + \frac{N_e e^2}{\varepsilon_o m_e} \frac{1}{(\omega_o^2 - \omega^2 - i\omega\gamma)}, \quad \text{(A.11)}$$

where γ is known as the damping frequency. Finally, it would be reasonable to generalize the matters by considering N as the number of atoms per unit volume, each oscillator having natural frequency ω_o so that the earlier expression becomes:

$$\varepsilon(\omega) = 1 + \frac{Ne^2}{m_e \varepsilon_o} \frac{f}{(\omega_o^2 - \omega^2 - i\omega\gamma)}, \quad \text{(A.12)}$$

where f is known as the oscillator strength. This model for dielectrics is known as the Lorentz model of damped harmonic oscillators.

A.2.2 *Metals: Drude model*

Imagine any conducting material as an assembly of driven and damped oscillators. Some of them correspond to free electrons with zero restoring force, while some others are bound to the atom as the case in dielectrics. If we assume that the average field which an electron moving within the conductor feels is just equal to the magnitude of the applied field, the general dispersion equation of a medium may be written as:

$$\varepsilon(\omega) = 1 + \frac{Ne^2}{\varepsilon_o m_e} \left[\frac{f_e}{(-\omega^2 - i\gamma_e \omega)} + \sum_j \frac{f_j}{\omega_{oj}^2 - \omega^2 - i\gamma_j \omega} \right], \quad \text{(A.13)}$$

The first term inside the bracket comes from the free electrons and N is same as is for the case of dielectrics. Each atom has f_e number of conduction electrons. The second term in the brackets comes from the bound electrons. With the help of this general expression (Eq. (A.13)), we can understand that how metals respond to incident light by neglecting the bound electron contribution and taking $f_e = 1$. Thus, we get:

$$\varepsilon(\omega) = 1 - \frac{Ne^2}{\varepsilon_o m_e} \frac{1}{(\omega^2 + i\gamma_e\omega)}. \tag{A.14}$$

In the expression (A.14), γ_e represents the damping frequency (or collision frequency) of free electrons and is represented as ω_{cb} for metals. Free electrons and positive ions within a metal may collectively be considered as a plasma with its density oscillating at a frequency ω_{pb}, called the plasma frequency, which may be given as equal to $(Ne^2/\varepsilon_0 m_e)^{1/2}$, and so:

$$\varepsilon(\omega) = 1 - \frac{\omega_{pb}^2}{\omega(\omega + i\omega_{cb})}, \tag{A.15}$$

Equation (A.15) is known as the Drude model of metal dielectric function. The plasma frequency acts as critical value below which, the refractive index $\left(=\sqrt{\varepsilon(\omega)}\right)$ of the metal is complex and the penetrating wave drops off exponentially from the boundary, i.e., the absorption is very significant. At frequencies above plasma frequency, refractive index is real and absorption is small. Thus, the absorption of a light wave in a metal is frequency dependent.

Appendix B

List of Constants

B.1 Sellmeier Relation

The Sellmeier relation for silica core of fiber is given as:

$$n_1(\lambda) = \sqrt{1 + \frac{a_1\lambda^2}{\lambda^2 - b_1^2} + \frac{a_2\lambda^2}{\lambda^2 - b_2^2} + \frac{a_3\lambda^2}{\lambda^2 - b_3^2}}. \tag{B.1}$$

Table B.1. Values of sellmeier coefficients.

Constants	Numeric value
a_1	0.696 1663
a_2	0.407 9426
a_3	0.897 4794
b_1	0.068 4043
b_2	0.116 2414
b_3	9.896 161

B.2 Plasma and Collision Wavelengths for Plasmonic Metals

The Drude model is given as:

$$\varepsilon_m(\lambda) = 1 - \frac{\lambda^2\lambda_c}{\lambda_p^2(\lambda_c + i\lambda)}. \tag{B.2}$$

Table B.2. Plasma and collision wavelengths of plasmonic metals and metal oxide.

Metal	Plasma Wavelength (λ_p) in meter	Collision Wavelength (λ_p) in meter
Gold (Au)	1.6826×10^{-7}	8.9342×10^{-6}
Silver (Ag)	1.4541×10^{-7}	17.6140×10^{-6}
Copper (Cu)	1.3617×10^{-7}	40.852×10^{-6}
Aluminum (Al)	1.0657×10^{-7}	2.4511×10^{-5}
Indium Tin Oxide (ITO)	5.6497×10^{-7}	11.21076×10^{-6}

B.3 Dispersion Relations of Various Dielectric Materials and Metal Oxides

The dispersion relation for the silicon (Si) dielectric is given as:

$$n_{Si} = A + A_1 e^{-\lambda/t_1} + A_2 e^{-\lambda/t_2}. \tag{B.3}$$

Table B.3. Coefficients for Si.

Constants	Numeric values
A	3:44904
A_1	271:88813
A_2	3:39538
t_1	0:05304
t_2	0:30384.

The dispersion relation for the zinc oxide (ZnO) is given as:

$$n_{ZnO} = \sqrt{m + \frac{m_1\lambda^2}{\lambda^2 - n_1^2} + m_2\lambda^2}. \tag{B.4}$$

Table B.4. Coefficients for ZnO.

Constants	Numeric values
m	2.81418
m_1	0.87968
m_2	−0.00711
n_1	0.3042

The dispersion relation for tin dioxide or stannic oxide (SnO_2) is given as:

$$n_{SnO_2} = q_1\lambda^5 + q_2\lambda^4 + q_3\lambda^3 + q_4\lambda^2 + q_5\lambda + q_6. \tag{B.5}$$

Table B.5. Coefficients for SnO_2.

Constants	Numeric values
q_1	178.76
q_2	−511.74
q_3	570.59
q_4	−306.11
q_5	77.206
q_6	−5.1211

The dispersion relation for the titanium dioxide (TiO_2) is given as:

$$n_{TiO_2} = \sqrt{p + \frac{p_1}{\lambda^2 - p_2}}. \tag{B.6}$$

Table B.6. Coefficients for TiO_2.

Constants	Numeric values
p	5.913
p_1	0.2441
p_2	0.0843

Index